KU-350-473

THE POWER OF BAD

Also by John Tierney and Roy F. Baumeister

Willpower: Rediscovering the Greatest Human Strength

Also by John Tierney

God Is My Broker: A Monk-Tycoon Reveals the 7½ Laws of Spiritual and Financial Growth (with Christopher Buckley)

The Best-Case Scenario Handbook

Also by Roy F. Baumeister

Meanings of Life

Evil: Inside Human Violence and Cruelty

The Cultural Animal: Human Nature, Meaning, and Social Life

Is There Anything Good About Men? How Cultures Flourish by Exploiting Men

Homo Prospectus (with Martin E. P. Seligman, Peter Railton, and Chandra Sripada)

Losing Control: How and Why People Fail at Self-Regulation (with Todd Heatherton and Dianne M. Tice)

Breaking Hearts: The Two Sides of Unrequited Love (with Sara R. Wotman)

The Social Dimension of Sex (with Dianne M. Tice)

Your Own Worst Enemy: Understanding the Paradox of Self-Defeating Behavior (with Steven Berglas)

Escaping the Self: Alcohol, Spirituality, Masochism, and Other Flights from the Burden of Selfhood

Masochism and the Self

Identity: Cultural Change and the Struggle for Self

THE POWER OF BAD

And How to Overcome It

John Tierney

AND

Roy F. Baumeister

ALLEN LANE
an imprint of
PENGUIN BOOKS

ALLEN LANE

UK | USA | Canada | Ireland | Australia
India | New Zealand | South Africa

Allen Lane is part of the Penguin Random House group of companies
whose addresses can be found at global.penguinrandomhouse.com

First published in the USA by the Penguin Press 2019
First Published in Great Britain by Allen Lane 2019
001

Printed and bound in Great Britain by Clays Ltd, Elcograf S.p.A.

A CIP catalogue record for this book is available from the British Library

Hardback ISBN: 978–1–846–14759–3
Trade paperback ISBN: 978–1–846–14760–9

www.greenpenguin.co.uk

Contents

THE POWER OF BAD

PROLOGUE

The Negativity Effect

Take the bad with the good, we stoically tell ourselves. But that's not how the brain works. Our minds and lives are skewed by a fundamental imbalance that is just now becoming clear to scientists: Bad is stronger than good.

This power of bad goes by several names in the academic literature: the negativity bias, negativity dominance, or simply the negativity effect. By any name, it means the universal tendency for negative events and emotions to affect us more strongly than positive ones. We're devastated by a word of criticism but unmoved by a shower of praise. We see the hostile face in the crowd and miss all the friendly smiles. The negativity effect sounds depressing—and it often is—but it doesn't have to be the end of the story. Bad is stronger, but good can prevail if we know what we're up against.

By recognizing the negativity effect and overriding our innate responses, we can break destructive patterns, think more effectively about the future, and exploit the remarkable benefits of this bias. Bad luck, bad news, and bad feelings create powerful incentives—the most powerful, in fact—to make us stronger, smarter, and kinder. Bad can be put to perfectly good uses, but only if the rational brain understands its irrational impact. Beating bad, especially in a digital world that magnifies its power, takes wisdom and effort.

The negativity effect is a simple principle with not-so-simple consequences. When we don't appreciate the power of bad to warp our judgment, we make terrible decisions. Our negativity bias explains things great and small: how countries blunder into disastrous wars, why neighbors feud and couples divorce, how economies stagnate, why applicants flub job interviews, how schools are failing students, why football coaches punt much too often. The negativity effect destroys reputations and bankrupts companies. It promotes tribalism and xenophobia. It spreads bogus scares that have left Americans angrier and Zambians hungrier. It ignites moral panics among both liberals and conservatives. It poisons politics and elects demagogues.

Bad is universally powerful, but it is not invincible. You are most affected by the negativity effect during your younger years, when you most need to learn from failures and criticism. As you age, the need to learn diminishes while perspective increases. Old people tend to be more contented than young people because their emotions and judgments aren't as skewed by problems and setbacks. They counteract the power of bad by appreciating pleasures each day and recalling happy moments instead of dwelling on past miseries. Their lives may not seem better by objective standards (particularly if they have health problems), but they feel better and can make sounder decisions because they can afford to ignore unpleasant learning opportunities and focus on what brings joy.

That's the sort of wisdom we're promoting in this book. We'll explain how to use the power of bad when it's beneficial and overcome it when it's not. Thanks to a recent surge of studies of the negativity effect, researchers have identified strategies for coping with it. Evolution has left us vulnerable to bad, which rules a primal region of the brain in all animals, but it also has equipped the more sophisticated regions of the human brain with natural cognitive tools for withstanding bad and employing it constructively. Today these tools are more essential than ever because there are so many more skilled purveyors of fear and vitriol—the merchants of bad, as we call them, who have prospered financially and politically by frightening the public and fomenting hatred.

We'll show how to deploy the rational brain to keep bad at bay in both private and public life—in love and friendships, at home and school and work, in business and politics and government. Above all, we want to show how good can win in the end. It is not as immediately powerful and emotionally compelling as bad, but good can prevail through persistence, intelligence, and force of numbers.

By learning how the negativity bias affects you and everyone else, you see the world more realistically—and less fearfully. You can consciously override the impulses that cause crippling insecurities, panic attacks, and phobias like the fear of heights or public speaking. A phobia is a discrete illustration of the power of bad: an exaggerated reaction to the possibility of something going wrong, an irrational impulse that prevents you from enjoying life to its fullest. Phobias can be overcome, and so can more generalized problems once you understand the negativity effect.

Instead of despairing at a setback, you can look for ways to benefit from it. Instead of striving to be a perfect parent or partner, you can concentrate on avoiding the basic mistakes that matter much more than your good deeds. In any relationship, you can learn how to stop fights before they begin, or at least prevent them from spiraling out of control, by recognizing how easily a small affront can be misinterpreted and exaggerated, especially when romantic partners are trying to make sense of each other. At work, you can avoid the pitfalls that ruin careers and doom enterprises.

The upside of bad is its power to sharpen the mind and energize the will. By understanding the impact of painful feedback, you become better at dealing with criticism—at absorbing the useful lessons without being demoralized. You also become better at dispensing criticism, a rare skill. Most people, including supposed experts, don't know how to deliver bad news because they don't realize how it's received. When doctors ineptly deliver a grim diagnosis, they compound the patients' grief and confusion. When students or employees are evaluated, many teachers or supervisors deliver critiques that serve mainly to dishearten, while others just duck the problems by giving everyone good grades and evaluations. They could do their jobs more effectively with techniques that have been tested recently in schools, offices, and factories.

Criticism and penalties, when administered deftly, spur much faster progress than the everybody-gets-a-trophy approach. They inspire people to learn from their mistakes instead of continuing to jeopardize their careers and their relationships. Criticism and penalties teach people how to improve themselves and get along with others, whether they're collaborating at work, juggling family responsibilities, or trying to keep romance alive.

Properly understood, the power of bad can bring out the best in anyone.

THE NEGATIVITY EFFECT IS A FUNDAMENTAL ASPECT OF PSYCHOLOGY and an important truth about life, yet it was discovered only recently, and quite unexpectedly. Roy Baumeister's research began, as usual, with a vague question, the sort that's no longer fashionable among his fellow researchers in psychology. As an undergraduate he had wanted to become a philosopher contemplating broad questions about life, but his parents considered that too impractical a career to justify Princeton's tuition, so he compromised by going into social psychology.

Once he became a professor, first at Case Western Reserve University and then at Florida State and the University of Queensland, Baumeister did his share of highly specialized research and experiments, the kind of work favored by today's journals and tenure committees. He became known for his work on self-control, social rejection, aggression, and other topics. But he also took on questions far beyond his specialties. Why is there evil? What is the self? What shapes human nature? What is the meaning of life? He answered each one in a book by surveying the literature in psychology and other disciplines to spot patterns unseen by the specialists.

In the 1990s he became intrigued by a couple of patterns in good and bad events. Psychologists studying people's reactions found that a bad first impression had a much greater impact than a good first impression, and experiments by behavioral economists showed that a financial loss loomed much larger than a corresponding financial gain. What gave bad its greater power? When and how could it be counteracted?

To investigate, Baumeister started by looking for situations in which bad events didn't have such a strong impact. It was a logical enough approach: To understand the source of something's strength, look for examples of its weakness. To find out what's supporting a roof, look for spots where it's sagging. Baumeister and his colleagues proposed to "identify several contrary patterns" that would enable them to "develop an elaborate, complex, and nuanced theory about when bad is stronger versus when good is stronger."

But they couldn't. To their surprise, despite scouring the research literature in psychology, sociology, economics, anthropology, and other disciplines, they couldn't find compelling counterexamples of good being stronger. Studies showed that bad health or bad parenting makes much more difference than good health or good parenting. The impact of bad events lasts longer than that of good events. A negative image (a photograph of a dead animal) stimulates more electrical activity in the brain than does a positive image (a bowl of chocolate ice cream). The pain of criticism is much stronger than the pleasure of praise. Penalties motivate students and workers more than rewards. A bad reputation is much easier to acquire and tougher to lose than a good reputation. The survey of the research literature showed bad to be relentlessly stronger than good. Almost by chance, the psychologists had discovered a major phenomenon, one that extended into so many different fields that the overall pattern had escaped notice.

While he was writing up the results, Baumeister happened to visit the University of Pennsylvania and present his findings. A professor in the audience, Paul Rozin, came up afterward and told him he was working on a similar project, although from a different approach. Rozin was already well known for his highly creative research into neglected topics, including magical thinking and disgust.

In a memorable set of experiments, he showed how little it took to contaminate something good. When a sterilized, dead cockroach was dunked into a glass of apple juice and then quickly removed, most people refused to take a sip. (The notable exception: little boys, who seemed incapable of being grossed out.) Most adults became unwilling to drink any apple juice at all, not even when it was freshly poured from a new carton into a clean glass. The slightest touch with a disgusting bug could make any food suddenly seem inedible.

But suppose an experimenter put a luscious piece of molten chocolate cake on top of a plateful of sterilized cockroaches. Would that make you willing to eat the bugs? Can you imagine any food so good that merely touching it to the plate would render the cockroaches edible? No, because there is no "anti-cockroach." Rozin's study of disgust and contagion confirmed an old Russian saying: "A spoonful of tar can spoil a barrel of honey, but a spoonful of honey does nothing for a barrel of tar."

As Rozin pondered this asymmetry, he saw that this negativity bias applied to a wide range of phenomena. In many religious traditions, a person can be damned by a single transgression or possessed by a demon in an instant, but it takes decades of good works and dedication to become holy. In the Hindu caste system, a Brahman is contaminated by eating food prepared by someone from a lower caste, but an untouchable does not become any purer by eating food prepared by a Brahman.

A few linguistic peculiarities also struck both Baumeister and Rozin. Psychologists generally describe emotional states with pairs of opposites: *happy* or *sad, relaxed* or *anxious, pleased* or *angry, friendly* or *hostile, optimistic* or *pessimistic.* But when Baumeister surveyed psychological research into good and bad events, he noticed that something was missing. Psychologists have long known that people can be scarred for years by a single event. The term for it is *trauma,* but what is the opposite? What word would describe a positive emotional state that lingers for decades in response to a single event?

There is no opposite of *trauma,* because no single good event has such a lasting impact. You can consciously recall happy moments from your past, but the ones that suddenly pop into your head uninvited—the involuntary memories, as psychologists call them—tend to be unhappy. Bad moments create unconscious feelings that don't go away. Fifty years after World War II, when researchers compared American veterans who'd fought in the Pacific with those who'd fought in Europe, there was a distinct difference in tastes: The Pacific veterans still avoided Asian food. One bad sexual experience can haunt a person for life, but the most blissful tryst will become a hazy memory. One infidelity can destroy a marriage, but no act of devotion can permanently bond a couple. One moment of parental neglect can lead to decades of angst and therapy, but no one spends adulthood fixated on that wonderful day at the zoo.

Rozin noticed some other singular bad words. For instance, there was no single word meaning the opposite of *murderer.* When the researchers tested this notion by asking people to name one, there was no consensus. Some people couldn't think of any word; others suggested words that were not quite right, like *savior* (a broader term typically used for spiritual redemption and other kinds of rescue) and *lifesaver* (which brings to mind something on a ship's deck). Previous researchers had studied languages around the world and found a negativity bias in the distribution of words: There are more synonyms for a bad concept like pain than for its opposite, pleasure. But for *murderer* there is no opposite. The Penn researchers looked for other such "unique nouns," either good or bad, and came up with just a handful—all of them bad.

They could find synonyms for *sympathy* (like *compassion* and *pity*) but no single word to connote empathizing with someone's good fortune. There was a word for an unexpected negative event, *accident*, and also for the chance that something bad could occur, *risk*, but most people couldn't think of an opposite for either one. (*Serendipity* is a possibility, but it apparently wasn't familiar to most people.) Nor could most people name an antonym for *disgust*. It was the same story when the researchers looked for versions of these words in twenty other languages, including the most widely spoken tongues as well as less common ones like Icelandic and Ibo. The results demonstrated an extreme version of the negativity bias: Sometimes bad is so much stronger that people don't even try contrasting it with good.

By the time they finished comparing notes, Baumeister and Rozin realized they had independently recognized the same principle, and they coordinated the publication of their papers in 2001. Both are now among the most cited papers in the social-science literature. They've inspired psychologists and a wide range of other researchers to conduct hundreds of studies of the negativity bias, discovering it in new places, analyzing its effects, and testing countermeasures. With this book we want to start sharing this growing body of research, which has deepened our understanding of the negativity effect while also confirming the original papers.

Rozin's paper, coauthored with his Penn colleague Edward Royzman, was titled "Negativity Bias, Negativity Dominance, and Contagion." They concluded that "negative events are more salient, potent, dominant in combinations, and generally efficacious than positive events." Baumeister's paper was titled simply "Bad Is Stronger Than Good." It was cowritten with two colleagues at Case Western, Ellen Bratslavsky and Kathleen Vohs, and Catrin Finkenauer of the Free University of Amsterdam. After surveying the evidence, they concluded: "The greater power of bad events over good ones is found in everyday events, major life events (e.g., trauma), close relationship outcomes, social network patterns, interpersonal interactions, and learning processes."

Baumeister and his coauthors noted that their own profession had been skewed for a century by the power of bad. Psychology journals and textbooks had devoted more than twice as much space to analyzing problems than to identifying sources of happiness and well-being. Why? "One hypothesis might be that psychologists are pessimistic misanthropes or sadists who derive perverse satisfaction from studying human suffering and failure." But a better explanation for Baumeister's team was the pressure on researchers in this young science to come up with statistically significant results: "They needed to study the strongest possible effects in order for the truth to shine through the gloom of error variance and to register on their measures. If bad is stronger than good, then early psychologists would inevitably gravitate toward studying the negative and troubled side of human life."

Researchers had been following their own version of the Anna Karenina principle, named for Tolstoy's famous observation that all happy families are alike but each unhappy family is unhappy in its own way. It was much easier to distinguish and measure the problems of unhappy people, so psychologists had started with them. The research was further distorted when it reached the public, because it was filtered through journalists eager for news with the most immediate impact—which, of course, meant bad news. So they wrote lots of stories about the toll of traumas and psychoses and depression, but precious little about the mind's resilience and capacity for happiness.

Post-traumatic stress syndrome became common knowledge but not the concept of post-traumatic growth, which is actually far more prevalent. Most people who undergo trauma ultimately feel that the experience has made them stronger, wiser, more mature, more tolerant and understanding, or in some other way a better person. The influential psychologist Martin Seligman has often lamented that so much attention is lavished on post-traumatic stress syndrome rather than post-traumatic growth because it causes people to mistakenly expect that bad events will have mainly negative effects. After being exposed to a terrifying event, at least 80 percent of people do not experience post-traumatic stress syndrome. Even though a single bad event is more powerful than a good event, over time people respond in so many constructive ways that they typically emerge more capable than ever of confronting life's challenges. Bad can make us stronger in the end.

Psychologists and journalists were so busy accentuating the negative that they missed the larger truth about human resilience. It was only after recognizing the negativity bias in their own field that psychologists began compensating for it by studying ways to foster hardiness, growth, and well-being rather than merely alleviate misery. To do that, they began to take a closer look at the power of bad, as did researchers in other disciplines. Cognitive scientists found new ways to counteract its effects in treating anxiety and other disorders, and to use its power to promote faster learning. Economists began to see how it could improve workers' productivity. Sociologists studying religion saw how the power of bad inspires virtuous behavior, and why hell is such a common belief in religions that spread quickly. The Christian doctrine of original sin—that humanity is doomed to perpetual suffering by the sin of Adam and Eve—may seem harsh, just as it seems unfair that the hero in ancient Greek tragedies is doomed by a single tragic flaw. But these beliefs happen to jibe with a basic element of human psychology and evolution.

To survive, life has to win every day. Death has to win just once. A small error or miscalculation can wipe out all the successes. The negativity bias is *adaptive*, the term biologists use for a trait that improves the odds of survival for an individual or a group. On our ancestral savanna, the hunter-gatherers who survived were the ones who paid more attention to shunning poisonous berries than to savoring delicious ones. They were more alert to predatory lions than to tasty gazelles. Recognizing a friend's kindness usually wasn't a matter of life or death, but ignoring an enemy's animosity could be fatal. At the group level, survival depended on what researchers call the chain principle (based on the cliché about a chain being only as good as its weakest link): The clan's safety could not be ensured by a single good person, but they could all be poisoned if one careless cook served tubers without eliminating the toxins. A single traitor could betray everyone to a hostile clan.

One mistake can still kill you. One enemy can still make your life miserable. One loss can erase many previous gains. Paying extra attention to threats still makes evolutionary sense. But our fine-tuned sense of bad can be debilitating, and what worked for hunter-gatherers doesn't always work for us. The urge to load up on fattening calories was useful in lean times on the savanna, but it can lead to obesity and ill health when junk-food merchants tempt you all day long. Bad today also has its merchants, and they use the media just as skillfully as junk-food marketers.

That's why the modern world seems so dangerous. Terrorism is a creation of the media age. Randomly murdering a few innocent civilians was strategically pointless until the late nineteenth century. Only then, as the telegraph and cheap printing presses began quickly spreading news, did terrorists discover the power of a single horrendous act. The quest for fear accelerated with broadcast news and has been in hyperdrive since cable channels and websites and social media started competing for audiences 24/7. They tap into primal emotions by hyping threats from nature, technology, foreigners, and political opponents. The election of Donald Trump has been a ratings bonanza because it has brought out the worst on both sides, so that rarely a week goes by without some new warning that Western civilization is doomed.

All day long, the power of bad governs our moods and guides our decisions. It drives news and shapes public discourse as it's exploited by journalists, politicians, marketers, bloggers, social-media vipers, Internet trolls, and anyone else seeking attention on our screens. The past quarter century has been extraordinarily peaceful by historical standards, but people have witnessed more battles and bloodshed than ever before. The rate of violent crime in America has plummeted, but most people think it has gone up because they see it so often in the media. The steady diet of bad news makes people feel helpless. They start catastrophizing their personal worries and despairing at the state of the world.

As life expectancy increases, we use our extra time to click on headlines like "Why Your Diet Is Killing You." No matter how happy your home life, you're assailed with listicles of the seven signs your partner is cheating and the five tips to prevent your child from being abducted. No matter how virtuously you live, clickbaiters will find a way to frighten you. You're not safe even in the web's realm of adorable-animal videos, not when there's an algorithm alerting you to articles aimed at pet owners: "Would Your Dog Eat You If You Died? Get the Facts."

Until we learn how to override the disproportionate impact of bad, it distorts our emotions and our view of the world. It has made the luckiest people in history feel cursed. For thousands of years, the normal human lot was a short life of hard toil on a farm. In 1950 most people in the world subsisted on less than $1 per day and didn't know how to read, but today the rates of extreme poverty and youth illiteracy are below 10 percent and still falling. We are richer, healthier, freer, and safer than our ancestors could have ever hoped to be, yet we don't enjoy our blessings. We prefer to heed—and vote for—the voices telling us the world is going to hell. Instead of seizing opportunities and expanding our horizons, we seethe at injustices and dread disasters—and all too often respond by making things worse.

The negativity bias causes us to pay special attention to external threats and thus exaggerate those dangers, but we're prone to a different bias when looking inward. We typically exaggerate our virtues, and our capacity for self-delusion can be astonishing. When prisoners serving time for assault, robbery, fraud, and other crimes were asked to compare themselves with the general population, they rated themselves as being more moral and honest, and also more compassionate and self-controlled. There was just one quality in which they didn't outshine the rest of society. When it came to being law-abiding, these convicted criminals modestly rated themselves as only average.

We're all prone to overestimate our abilities as well as our power to control our destiny. People have a false sense of security on the highway because they consider themselves above-average drivers and expect their skill to protect them, even though many accidents are caused by factors beyond their control. Similarly, when asked how long it will take to complete a project, people typically underestimate the time because they're too confident in themselves and don't allow for delays beyond their control. This "optimism bias" causes people to underestimate the risk of some types of negative events in their own lives. They're fully aware that something bad can happen—in fact, they often have an unrealistically high expectation it will happen—but they tell themselves it will happen to someone else.

Over and over, this toxic combination of fear and overconfidence leads to disaster. Political scientists have used it to understand some of the most puzzling mistakes of modern history, starting with the carnage of World War I. Why was Germany so eager for a war that proved so futile? Before the war, Germany was the greatest economic and military power in Europe, so strong that its neighbors would have been foolhardy to attack. Yet German leaders obsessed over any signs of hostility from other nations. In 1912 the German chancellor wondered whether it would be worthwhile to plant trees on his estate because he assumed "the Russians would be here in a few years in any case." While historians have struggled to come up with rational reasons for this paranoia, the best explanation lies in the psychological literature, according to the political scientists Dominic Johnson and Dominic Tierney (no relation).

They've recently drawn on the work of Baumeister and Rozin to explain the fears that led German leaders into World War I and also spurred the American decision to invade Iraq in 2003. Like the Germans, the Americans overestimated the threat from their enemy, mistakenly believing Saddam Hussein had weapons of mass destruction. And like the Germans, who expected a quick victory, the Americans suffered from optimism bias when estimating their own ability to replace Saddam with a stable democratic government, so in their zeal to eliminate an imagined danger they created a real one by fostering the chaos that enabled ISIS and other jihadist groups to thrive.

These wars are examples of what we call the Crisis Crisis: the never-ending series of hyped threats leading to actions that leave everyone worse off. The United States is the greatest military power in history, but politicians want us to believe the nation is mortally threatened by Iran and North Korea. The safer that streets become, the harder the media search for new menaces, like the imagined waves of violence by illegal immigrants (who are probably less likely than natives to commit crimes) or the supposedly rampant "stranger danger" from homicidal abductors of children (a risk that's much lower than the risk of being killed by lightning). If it's a slow news day, there are always future doomsdays to fear—a virus wiping out humanity, a takeover of the world by robots, a global environmental collapse. Apocalyptic predictions have become so common that when a national sample of preteen children in America were asked what the planet would be like when they grew up, one in three of the children feared that Earth would no longer exist.

The precise term for the adults scaring these children is *availability entrepreneurs.* They're the journalists, activists, academics, trial lawyers, and politicians who capitalize on the human tendency to gauge a danger according to how many examples are readily available in our minds. The number of people killed worldwide by al-Qaeda and ISIS and their allies in the past two decades is smaller than the number of Americans who died in their bathtubs, but we see the victims of terrorism over and over on our screens. The result is a self-perpetuating process that Timur Kuran and Cass Sunstein have termed an availability cascade: News coverage of a danger creates public fear, inspiring further coverage and more fear, which is why 40 percent of Americans worry that they or a family member will die in a terrorist attack. Meanwhile, lacking the sensational media coverage of their death toll, bathtubs fail to inspire dread, and millions of Americans climb in and out without pausing to fear for their lives.

WE WANT TO COUNTERACT THE CASCADES OF FEAR THAT PRODUCE needless personal angst and destructive public policies. We hope to start a different sort of cascade. We fully expect Earth to be around when today's children grow up, and we'd like them and their parents to share our optimism. Life is no longer "nasty, brutish and short," as Thomas Hobbes described the plight of early humans, but psychologists have found that even the most affluent and long-lived people still see it that way. When the researchers asked adults in the United States, Canada, and India whether life is long or short, and whether it's easy or hard, the North Americans were no more sanguine than the Indians despite their statistical advantages in life expectancy and income. Barely one in eight of the North Americans considered life to be both long and easy, while a majority considered it both short and hard. The optimists, not surprisingly, were significantly happier than the pessimists, and they were also more public-spirited—more likely to vote, to donate to charity, and to do volunteer work in their community.

How to increase the ranks of optimists? We certainly don't expect to eliminate the negativity effect, but we hope to show you how not to be ruled by it. First, we'll explore its power—how much stronger bad is than good, how it operates in the brain, how it distorts your perceptions of people and risks, and how you can minimize those distortions. In the middle of the book, we'll discuss how to use the power of bad for positive purposes, and how to deal with the particular challenges of the negativity effect in business and the online world. Then we'll look at the innate human strengths and conscious strategies that can be marshaled against the modern barrage of bad.

Humans are unique among animals in our ability to control—or at least recognize—the negativity bias. Other creatures have innate aversions to danger as well as mechanisms enabling them to learn to dislike things, often quite rapidly, but humans have a singular ability to overcome aversion. We often come to love activities that initially terrified us, like watching a horror movie or riding a roller coaster. We recoil at our first taste of coffee or garlic or hot peppers, but then we learn to appreciate the experience. The fear of falling is innate—infants display it before they can speak—but some people become devotees of skydiving and bungee jumping.

We can gain the perspective to see that there's much more to celebrate than mourn in our lives and in the world, and to use that knowledge to make things still better. We can flourish despite the power of bad, but we must learn how, starting with the most basic strategy: Know the enemy.

CHAPTER 1

How Bad Is Bad?

Enlisting the Rational Mind

Early in his career, long before he published anything on the negativity effect, Baumeister conducted what might charitably be called a pilot study. The study sample consisted of himself. He was in a relationship with a partner who was brilliant, charming, and loving—most of the time. But sometimes she was provoked into screaming rages that left him despondent and confused. He'd never heard his parents raise their voices at each other, and he'd never been involved with anyone so volatile. She could be angered by what seemed to him innocent mistakes, like dripping water on the bathroom floor during a shower or forgetting to turn off the iron when he finished pressing his shirts. She once got so upset that she smashed a dish by hurling it at the kitchen wall. He knew he had his failings, and he realized relationships took work, but these confrontations were tough to endure.

Afterward, though, she'd be genuinely remorseful. She'd apologize, accepting the blame while calmly explaining why he'd upset her and what he could do differently the next time. His doubts would vanish as he listened to her, seeing her brilliance and charm shining through once again. He'd think back on their early times together, those thrilling moments of discovering a soul mate, and he knew he still loved her. They'd passionately reconcile, promising each other to try harder. She'd work on her temper, and he'd try to be a more considerate partner. Good times would return, but before long things would deteriorate. When he responded as she had suggested he do to help her, she discounted it as a ploy and remained furious at him, leaving him determined to end the relationship. But the next morning his hopes would revive again.

He could see that bad and good are not necessarily opposites. They can exist side by side, in separate domains. He knew a lover was supposed to follow his heart, but which heart on which day? On the bad days he desperately wanted to be free, but he didn't want to be alone, either. He'd grown up in a family where love seemed to be more of an obligation than a joy, so falling in love had been a world-changing discovery for him. Was it reckless to walk away from a relationship that brought bursts of such intense happiness? If he'd learned anything from his psychology classes, it was how easily the human mind could be whipsawed by emotions. He didn't trust his feelings, not when they swung so violently between bliss and despair. He wanted some way to invite his rational mind back into the conversation.

We all know the feeling. You're trying to assess something—a romance, a job, a friendship, a project—and you see the pluses and the minuses. You want to follow through and honor your commitment, but when is the cost too high? Your gut may be telling you to get out, but since bad is more viscerally powerful than good, your gut isn't necessarily reliable. In a heated or difficult moment, bad will loom larger. To properly weigh bad against good, you need to engage your rational mind—System 2, as the psychologist Daniel Kahneman calls the logical, slower-moving part of the brain. It takes more mental effort in the short run than going with your gut—Kahneman's System 1, the instinctive and emotional part that's quickly swayed by the power of bad—but in the long run it can spare you both energy and anguish.

In his romantic quandary, the young Baumeister fell back on the classic strategy of confused social scientists: collect data. He devised a crude binary measure. Each evening, he would look back on the day, ask himself if he was glad to be in the relationship, and mark a yes or no answer in his notebook. He also established a couple of limits. If it turned out that the bad days outnumbered the good days, he told himself, that would be a clear reason to break up. If there were at least four good days for every bad day, that would be a reason to stay together. In between would be, well, in between. He realized these parameters were arbitrary, but he was desperate for some kind of clarity.

After several months of record keeping, he could see that the ratio was remaining fairly steady—and that there was still no clarity. They had two good days for every bad day, a ratio midway between his limits. What to do? He could see that the good days outnumbered the bad days by a sizable margin, yet he was feeling more miserable than ever on the bad days. He decided to break up with her, a decision based on his gut rather than his data, but eventually other scientists provided a rationale for his decision—and also for his research method.

With those daily entries in his notebook, he had happened on a concept that would later be termed the *positivity ratio*, which is the number of good events for every bad event. This simple ratio can't measure the full complexity of love or life, but it's a valuable tool for understanding the negativity effect. It enables researchers to measure bad and gauge its impact. It gives therapists and counselors a way to diagnose problems and assess progress.

Perhaps most important, it gives all of us a way to deal with the power of bad: to use the rational part of the brain to understand and override the debilitating fears and anxieties that constrict lives, warp decisions, and ruin relationships. To figure out how well a person or a couple or a group is doing, and to overcome the negativity bias, you need a method of weighing the positive against the negative to determine their relative strength. You need to ask: Just how bad is bad?

Let Us Count the Ways

One of the pioneering researchers into the positivity ratio was Robert Schwartz, a clinical psychologist who wondered how much help he and his fellow therapists were providing to patients. He wanted a more precise measure of progress than "the client was less depressed after treatment." Over several decades starting in the 1980s, he compared the number of positive and negative feelings reported by people undergoing psychotherapy. He found that severely depressed people tended to have twice as many negative feelings as positive feelings, and that this ratio could be improved by talk therapy and antidepressant drugs.

At the other extreme were people with positive feelings 90 percent of the time, who seemed dangerously unrealistic and prone to egotism, mania, and denial. Life isn't relentlessly happy, and the healthy person shows some reaction to the bad—but not too much. Schwartz concluded that people with equal numbers of positive and negative feelings tended to be "mildly dysfunctional," while those with "normal functioning" averaged about two and a half positive feelings for every negative one. The patients who achieved "optimal functioning" averaged a little over four positive feelings for every negative one. All of this may sound rather theoretical, but helping therapists measure feelings with more accuracy was an important step in treating afflictions like depression.

Other researchers have been unpoetically counting the ways in which people love each other—or don't. One simple method has been to count the number of times a couple had sex and the number of times they argued. Neither number by itself is revealing: Some happy couples have few arguments and little sex, and some have lots of fights and makeup sex. But the ratio between sex and arguing has turned out to be a reliable predictor of a marriage's prospects.

A more ambitious method is to count the way partners get along with each other. The psychologist Harris Friedman did an early study, in 1971, recording the number of positive and negative comments that husbands and wives made to each other while playing a stressful game that required them to cooperate. He found that the ratio of positive to negative comments during the game correlated with the couple's satisfaction with their marriage. In other studies, the psychologist John Gottman found that partners in a troubled relationship have an equal number of bad and good interactions, while those destined for long-term happiness have five times as many good interactions as bad ones.

This "Gottman ratio" of 5 to 1 has proved a useful standard for gauging quite different types of relationships. Some happy couples show little affection but flourish because they hardly ever fight; other successful couples fight more often but make up for it with lots of warmth and kindness. Informally, some researchers refer to this ratio as the "five fucks for every fight" rule. That's an oversimplification—there are many forms of affection besides sex—but it's a quick way of assessing the fundamental issue: Does the good significantly outweigh the bad? The Gottman ratio is a worthwhile target for couples, although it doesn't mean that bad is five times stronger than good.

Couples therapists advise the 5-to-1 ratio because it's well beyond the break-even point.

Behavioral economists have been studying positivity ratios using a conveniently simple measure: dollars. In experiments more than half a century ago, researchers quickly noticed that people would sometimes make irrational bets in their lust to make money—a finding that was not news to casino operators. But the experiments showed that people were even more irrational when there was a risk of losing money. This phenomenon later acquired a name, *loss aversion,* after research by the psychologists Daniel Kahneman and Amos Tversky. They found that most people are unwilling to make even-money bets on a coin toss. They won't risk losing $20 on a coin toss unless they're offered a potential gain of double that amount, $40. Why not? "Losses loom larger than gains," Kahneman and Tversky concluded.

But there's also another reason for those bettors' caution, as other researchers have recently found. It's not just that people hate losing money. They don't really believe that the coin toss is a 50-50 bet. They have a gut feeling that if they pick heads, then the coin is more likely to come up tails. That sounds crazy—it's certainly irrational—but it's common because of how people envision the future.

If, say, they're given identical weather forecasts for London and Madrid, a 10 percent chance of rain in each city, they'll typically think it's more likely to rain in London. It makes no sense mathematically, but the rain seems likelier in London because it's easier to envision wet weather in England than in Spain. The more familiar a scenario is—the more images of it that we've seen or imagined—the more likely it seems. That delusion can distort judgments about a coin toss. Experiments tracking gamblers' eye movements show that they pay more attention to a potential loss than to a gain. Since they spend more time thinking about the loss, they start to believe that it's more likely to occur, so they'll refuse an even-money bet. They'll demand odds of at least 2 to 1, and sometimes more, depending on how much money is involved and other factors. The economist Richard Thaler has found much bigger ratios by upping the emotional stakes.

The Rule of Four

We've seen that it takes somewhere between two and five good things to offset one bad thing. It's not surprising that the results at the lower end of the range come from studies involving money, because in those situations it's easier for the brain to overcome the negativity bias by concentrating on numbers instead of feelings. As much as it hurts to lose money, you can tell yourself that a loss of $100 is fully offset by a gain of $100. Gamblers routinely train themselves to play according to the mathematical odds rather than their gut instincts. Not everyone can do that, but in the experiments with money some people were no doubt able to balance gains and losses fairly rationally, and so the average ratio was about 2 to 1.

When money is not involved, though, it's much tougher to make precise comparisons, and most of the good and bad events in our lives don't evoke such rational responses. As we've mentioned, the ratio tends to be higher than 2 to 1 for people dealing with depressing thoughts or conflicts with their spouses. Research tracking workers' moods during the day shows that a setback has between two and five times as much emotional impact as a positive event. Emotions make us less rational, and therefore more susceptible to the power of bad.

Some of the most often cited measurements of emotional well-being have been made by the psychologist Barbara Fredrickson. After giving diagnostic tests to students when she was at the University of Michigan, she classified them into two categories: the flourishing and the languishing. The flourishing students felt a strong sense of purpose and control over life, accepted themselves, and got along well with others, as indicated by their test results. The languishing students were having more personal struggles and didn't feel as well integrated into the community.

Over the next month, both groups of students kept daily records of their highs and lows. Each evening they logged on to a website and rated how strongly, if at all, they had experienced different emotions during that day. The list included positive emotions (such as amusement, awe, joy, compassion, contentment, gratitude, and love) and negative ones (such as anger, contempt, sadness, embarrassment, guilt, and fear). When Fredrickson added up all the daily reports, she found that the languishing students had more positive than negative emotions, but the overall positivity ratio was only about 2 to 1. For the flourishing students, the overall positivity ratio was slightly more than 3 to 1.

The finding attracted attention because it pointed to the wider benefits of positive psychology. Fredrickson and other researchers had done earlier experiments in the laboratory demonstrating that when people are prompted with positive stimuli, they do better at creative tasks. They literally see the bigger picture. Their eyes sweep a wider field of vision instead of just focusing on what's straight ahead, as they do when prompted with negative stimuli. Fredrickson developed what she called the "broaden and build" theory: Positive emotions broaden your perspective and enable you to build skills that help you flourish both personally and professionally. The theory became one of the most influential ideas in positive psychology, and the study with the students' diaries offered real-world confirmation as well as a method for gauging how well someone is doing.

Researchers kept finding similar positivity ratios when measuring other kinds of good and bad effects. One of the simplest measures—and our favorite, because it was what Baumeister did during his troubled romance—is to count the number of good and bad days. Some researchers have done this by asking people whether they're having a good day, a bad day, or a typical day, as the psychologist Randy Larsen did in studies that tracked people's daily moods over periods ranging from one to three months. Besides identifying which positive and negative emotions they'd experienced, people also singled out the best and worst event of the day, and rated how intensely each event had affected them. Larsen combined all these answers to classify each day as predominantly positive or negative. Overall, he found that the typical person had three good days for every bad day.

To do better than average, then, you'd want to have at least four good days for every bad one. That strikes us as a useful goal, and not just because it happens to be the same one that the young Baumeister picked when he was counting the good and bad days of his romantic relationship. Since he made that guess, researchers have repeatedly found that bad things are at least twice as powerful as good things, and generally at least three times as powerful when dealing with emotions and relationships rather than with dollars and cents. For the good to outweigh the bad, that means the positivity ratio should be at least 3 to 1, and preferably a little higher. So we'll suggest a guideline that we've taken to calling the Rule of Four: *It takes four good things to overcome one bad thing.*

We offer this as a rough gauge. We're not claiming to have discovered a universal constant like the speed of light or Avogadro's number. It's a rule of thumb, not a law of nature. It doesn't apply to every person in every situation, nor to every kind of good and bad event. Some forms of bad are incomparably powerful. As we noted earlier, there are a handful of negative words like *trauma* and *murderer* that have no opposite positive word.

But virtually all negative words do have antonyms, because we juxtapose most bad things against good things. We spend most of our time dealing with a mix of good and bad experiences and feelings. We assess ourselves and our prospects by weighing the two. The Rule of

Four can help you evaluate a relationship or a job by using the same day-counting technique employed by Baumeister. If you have four good days at work Monday through Thursday, that's probably enough in a typical week to make up for a bad Friday. Obviously, that 4-to-1 ratio won't be much comfort if on Friday you get fired, but that wouldn't be a typical week. The rule is relevant only when the events are comparable in magnitude, like ordinary successes and setbacks at work or displays of affection and hostility at home. If you and your partner are engaging in sex at least four times as often as you argue, the relationship looks pretty healthy. If the ratio is only 2 or 3 to 1, your prospects are less certain. And if the count is 1 to 1, that's not a tie. That's trouble.

When you start a self-improvement regimen, like resolving to exercise every day or eat a healthier diet, the Rule of Four can make a useful target. People often abandon New Year's resolutions because they set unrealistic goals for themselves and then give up after the first slip. Dieters routinely succumb to what nutrition researchers call the what-the-hell effect: *Now that I've broken the diet with that bowl of ice cream, I may as well finish off the carton.* Instead of demanding perfection and despairing when you fail, you could aim to stick to your regimen at least four days out of five. That goal may be too lax for some undertakings—to quit smoking, a strict cold-turkey policy is often essential—but you'll probably benefit most of the time by keeping the virtue-to-vice ratio at 4 to 1 or better.

Keep that ratio in mind when considering the impact of your actions. If you're late for one meeting, you won't redeem yourself by being early the next time. If you say or do something hurtful, don't expect to atone for it with one bit of goodwill. Whether you're evaluating a romantic partner or an employee, plan on at least four compliments to make up for one bit of criticism. (We'll have more to say later about how and when to blend the compliments.) You can't always take the rule literally, of course. It doesn't mean that you should send four orders of flowers to make up for one faux pas. But it does mean that one batch of flowers probably won't undo the damage. Throw in some other forms of reparation. Remembering that ratio can help you deal with mistakes in relationships, at work, and in the rest of your life.

The Rule of Four can also be a useful gauge for assessing how well a company or a product is faring. Successful companies typically have at least three satisfied customers for every unsatisfied one, whether those feelings are assessed in surveys or in online reviews. Among the millions of businesses on Yelp, there are three positive reviews (4 or 5 stars) for every negative review (1 or 2 stars). So if you want to do better than average, aim for at least a 4-to-1 positivity ratio and pay special attention to unhappy customers (we'll discuss them in chapter 7).

Similarly, it pays to make special efforts to deal with any bit of bad publicity. Correcting the problem isn't enough. The bad publicity needs to be swamped with lots more good publicity, the way that the Cadbury chocolate company's subsidiary in India dealt with a public-relations fiasco in 2003. During October, the peak month for chocolate sales because of the annual Hindu "festival of lights," Diwali, some customers in Mumbai reported finding worms in their Cadbury bars. The Indian government announced an investigation of the company's factories, prompting a wave of news stories that caused Cadbury sales to plunge. The investigation exonerated the factories, showing that the problem was improper storage by retailers unaffiliated with Cadbury, but the firm's executives realized that the good news wouldn't overcome the impact of the bad publicity.

The company launched a multifront offensive called Project Vishwas, or Project Trust, to counter images of worm-infested candy bars. It introduced foil-lined "purity-sealed" packaging for Cadbury chocolate and provided retailers with metal containers and coolers for proper storage. It publicized these innovations by holding press conferences and road shows across India, issuing video news releases, sponsoring a children's quiz show on television, and publishing newspaper ads in eleven languages. At enormous expense, Cadbury ran television commercials featuring a movie star, Amitabh Bachchan, whom a consumer survey had shown to be one of the two most credible people in India (the other was the prime minister). The commercials showed him visiting Cadbury's factories, inspecting its new packaging, and assuring his own granddaughter that the candy was safe. This multipronged barrage eventually overwhelmed the bad publicity, reviving Cadbury's sales and providing business students around the world with a case study of effective crisis management.

In discussing the Rule of Four, so far we've been guilty ourselves of the negativity bias by stressing how much attention you need to pay to your own mistakes and problems. But there's also a flip side to this rule, a positive lesson to keep in mind when managing your reaction to external problems: Remember that the negativity effect can distort your judgment, and that you can override the brain's irrational impulses.

Superstitions are based in large part on the negativity effect. If a couple of good things happened to you after a black cat crossed your path, you'd barely notice, but just one bad thing could make you permanently superstitious. While there are some positive superstitions, like believing a horseshoe or a rabbit's foot will bring good luck, most of them are negative, just as most tales of the supernatural involve scary events. Surveys of folklore and mythology around the world have found many more references to malevolent gods and demons than to helpful deities, angels, and fairy godmothers. In laboratory experiments, psychologists have found that we not only pay more attention to bad events but also are more likely to ascribe them to external forces. When a baseball team has a series of good seasons, it's attributed to their skill, but when there's a long dry spell, fans come up with explanations like the Curse of the Bambino or the

Curse of the Billy Goat (which were blamed for the losing streaks of the Boston Red Sox and the Chicago Cubs).

Even when we don't invoke the supernatural, we still ascribe too much power to isolated bad events, but we can compensate by keeping the Rule of Four in mind. When you feel devastated by an insult or a critique, remind yourself that this feeling could be due to your negativity bias rather than your inadequacy. Instead of obsessing about a snarky comment on social media, scroll down and reread four compliments. When you're furious at a friend for letting you down, force yourself to recall the times that same friend came through for you.

Be especially careful in making judgments about outside groups. Much of today's political polarization stems from the outrage generated by isolated acts that are exploited by the opposition. Before you draw any conclusions based on a horror story about an immigrant, think of four other immigrants you deal with every day. Before you write off Republicans as racists or Democrats as Marxists, think of the ones in your own family. Before you despair at the day's news, remember that journalists and politicians can't resist making false generalizations. They routinely portray a rare event as the norm instead of looking at the larger picture. They'll focus on one sensational murder instead of the trends showing most neighborhoods are safer. They'll treat one factory closing as a symptom of American industrial decline—and an excuse for protectionist trade policies—when in fact the nation's manufacturing output has been growing for decades (a trend most Americans aren't aware of).

You can't repeal the Rule of Four, because one bad event is going to have more visceral impact than a couple of good events, but you don't have to make long-term judgments based on those gut feelings. If one thing goes wrong, don't interpret it as a harbinger of inevitable doom, whether you're dealing with a personal setback or contemplating the state of the world. Whatever dismal event leads a newscast, on most days there are a lot more than four good things happening for every bad one. That's why life has been improving for most people in the world. And that's why, when you lose perspective and overreact to the bad, you're liable to make things worse.

Safety Junkies

Of all the forms of addiction, the most costly is the one that gets the least attention: an addiction to safety. We pay so much attention to bad things—reliving them, imagining them, avoiding them—that we let fear run our lives and become irrationally cautious. We're so focused on averting one obvious danger that we fail to foresee more subtle pitfalls. So we pass up opportunities for happiness and success in our personal lives, and we adopt public policies that leave us paradoxically less safe.

Consider how the American public and government responded to the plane hijackings on September 11, 2001. Until that morning, the best strategy during a hijacking was for the pilot to let the hijacker take control of the plane, because it was assumed that the hijacker wanted to land the plane safely rather than kill himself along with everyone else. That tragically mistaken assumption created a vulnerability that allowed hijackers to crash three planes into their targets on September 11. But the fourth plane failed to reach its target, because the hijackers were overpowered by passengers who'd gotten word of the earlier attacks and quickly realized that this new tactic required a new response: Don't let the hijackers pilot the plane.

From that moment on, no terrorist could count on this hijacking strategy to work again. Even if passengers weren't willing to intervene, a pilot could simply seal himself in the cockpit and ignore the hijackers' demands. The one bit of good news on September 11 was that this new terrorist threat to aviation was already obsolete.

But the horror of the bad news overwhelmed everyone's judgment. Fears of another hijacking prompted a ban on all airline flights for two days, giving the terrorists another propaganda victory and causing more economic disruption than the attack itself. When flights resumed, passengers were forbidden to board with nail clippers and scissors, as if anyone could ever use these to hijack another plane. In its rush to protect air travelers, Congress created a bloated bureaucracy to screen passengers, the Transportation Security Administration, that would become notorious for inefficiency and incompetence as it squandered more than $50 billion over the next decade. The TSA, known to frustrated travelers as "Thousands Standing Around," has repeatedly flunked tests of its ability to detect genuine threats like bombs in luggage. Its folly was obvious from the start to aviation-security experts, who warned Congress in 2001 that a centralized bureaucracy would be much less effective than the decentralized systems used in Israel and other countries experienced in fighting terrorism.

But in the aftermath of September 11, Congress was too panicked to act rationally, and so were the American people. In the ensuing year, millions of them avoided air travel and chose instead to drive to their destinations, resulting in what has been dubbed "9/11, Act II." Since driving is riskier than flying, it's been estimated that the switch from planes to cars resulted in an additional 1,600 deaths.

Safety addiction is a problem even when the stakes are much lower, and even for people who have all the expertise and incentives to make smart decisions. Any golfer knows that making a birdie on a hole (one stroke under par) will fully offset a bogey on another hole (one stroke over). Yet even the best professionals don't play as if they believe it. They consider it more important to avoid a bogey than to make a birdie. "You don't ever want to drop a shot," Tiger Woods once explained. "The psychological difference between dropping a shot and making a birdie, I just think it's bigger to make a par putt." But by yielding to the power of bad, Woods and his fellow pros are consigning themselves to worse scores and less prize money, as the economists Devin Pope and Maurice Schweitzer found by analyzing millions of their putts.

When these pros have a chance for a birdie, they putt differently than if they were trying to make par. Instead of going all out to sink the ball, they soften their stroke and are more likely to leave it just short of the hole, guaranteeing themselves an easy putt for par rather than risking the chance of overshooting the hole by so much that they end up taking two more putts for the dreaded bogey. While this strategy does help them avert a bogey, it more often costs them a birdie, so at the end of a tournament they're typically one shot worse off than if they'd gone for the birdies. Over the course of a season, the economists calculate, this fear of bogeys costs the leading pros more than $600,000 apiece in prize money.

Professional football coaches in America are even more extreme safety addicts. They continually study performance statistics looking for any slight advantage, anything to score more points and justify their multimillion-dollar salaries. Yet they go on making the same simple mistake week after week when faced with the fourth-and-short decision. On fourth down, when they need to gain just one or two yards to retain possession of the ball, they routinely refuse to go for it. Instead, they send in their kicker to punt to the other team, sacrificing a chance to keep the ball in return for improving their field position by about forty yards.

This trade-off was worthwhile in the old days of low-scoring football, when defenses dominated and teams had a hard time advancing. But maintaining this tradition makes no sense with today's pass-driven offenses that move so readily down the field. After crunching the numbers from thousands of games in the National Football League, analysts have repeatedly concluded that unless a team is trapped deep in its own territory, punting on fourth and short is a bad strategy because the improved field position is so much less valuable than the chance to keep the ball and go on to score.

Gregg Easterbrook, the Tuesday Morning Quarterback columnist, has calculated that the typical NFL team would win one additional game each season—often the difference between making the playoffs and being eliminated—if it took more chances on fourth down. He advises teams to go for it on fourth and short unless they're inside their own twenty-yard line. The number crunchers at the *New York Times'* Upshot figure that when a team needs to gain just one yard, it should go for it anywhere on the field beyond its own eight-yard line. While the recommendations may sound foolhardy, in these fourth-and-short situations the odds are very much in the team's favor: They can expect to succeed twice for each time they fail.

But a 2-to-1 ratio, as we've seen, isn't usually enough to overcome the negativity bias. A coach knows that if he goes for it and succeeds, he'll get a little credit for it, but the play probably won't make the highlight reel of the game. If the attempt fails and the other team goes on to score, it will loom large. The sportscasters will denounce him as reckless and warn that this "loss of momentum" could be the "turning point" in the game. If the team ends up losing a close game, that failed attempt on fourth down will be blamed for the loss and replayed endlessly afterward.

So it takes a brave coach to make the smart play. It takes someone with the iconoclastic streak—and job security—of Bill Belichick, the New England Patriots coach renowned for his Super Bowl victories and strategic brilliance. But even Belichick goes for it only occasionally on fourth and short, and his reputation doesn't protect him from being denounced by sportscasters and fans when the attempt fails. A few other coaches have looked at the numbers and said they plan to take more risks on fourth down, but they have a hard time following through when the moment comes.

A coach can try telling himself that the odds are in his favor, but he's got those images of past highlight reels in his head, so he can much more vividly imagine failure than success—and therefore, as in those coin-toss experiments, his gut overestimates the chance of failure. Even if he went into the game resolved to make the smart statistical play, when the moment comes he'll be reluctant. He'll look for an excuse to play it safe, telling himself that the statistics don't apply here because the other team's defense is so strong, or one of his linemen is injured, or his running back is having a bad day. He'll ignore the numbers, send in the punter, and breathe a sigh of relief. Whatever happens, no one will blame him for playing it safe. The power of bad has prevailed.

There is one football coach, though, who has figured out how to overcome it, and his strategies are worth applying off the field, too.

Just Go for It

When asked how he became the football coach who never punts, Kevin Kelley's first explanation involves his reading habits. He spends less time reading books on football than books on psychology and behavioral economics. He knows all about the principle of loss aversion.

He realizes that the power of bad can make everyone irrational, including himself. Just before he spoke with us, he'd been trying to plug a charging cable into his phone, and on his first try he'd inserted it upside down. "It feels like I do that just about every time," he said. "I know the odds must be fifty-fifty of plugging it in right, but I would swear I get it wrong ninety percent of the time. It's amazing how we all think so negatively."

To overcome that bias in his job, Kelley has trained himself to shift perspective. The training and execution weren't easy, but Kelley considered it a necessity when he became the coach at Pulaski Academy in Little Rock, Arkansas. He was supposed to turn the high-school team from an also-ran into a title contender, but he'd be competing against bigger schools with more talent to draw on, and there was no way to eliminate that disadvantage. He figured he had one psychological advantage over his rivals: "I don't care a whole lot what other people say as long as I think I'm doing the right thing." He was willing to try something unconventional even if it brought him criticism and boos.

Before the season began, he studied football statistics and forced himself to focus on the good things that could happen to a team that went for it on fourth down. It was his version of a strategy that had been tested in some of the coin-tossing experiments by psychologists. As we mentioned earlier, one reason for people's irrational belief that the odds are against them in a coin toss is that they spend more time contemplating the prospect of a loss rather than a gain. When researchers prodded people to spend equal amounts of time contemplating gains, they subsequently became more rational in their bets. Kelley achieved the same effect by contemplating the extra points he'd score if he consistently went for it on fourth down. The benefits were obvious on paper, but he knew that during a game he'd be vulnerable to those gut fears telling him to punt.

"I wanted to make as many decisions as possible before the game, before emotions take over," he explained. "In the small picture, it makes sense to play it safe on fourth down. The fans won't be mad and you won't have to face the media if you fail. You need to look at the big picture and see how many points you'll score in the long run."

So Kelley made a simple rule for himself: no punting. His rule allowed exceptions in only a few very specific situations, like the last few seconds of a half when the team was near its own goal line and there was no time to score even if it made the first down. During his first season, the team averaged one punt per game, and then he further tightened the rule so that the team has since averaged just one punt per season. He doesn't merely go for it on fourth and short. Even if it's fourth and thirty on his own one-yard line, he follows his no-punt rule.

Life is more complicated than football, but this technique can often work off the field, too: Take the decision out of your hands. Instead of trusting your feelings in the moment, make a rule beforehand. A simple rule—if I'm in *x* situation, I will do *y*—can help you avoid safety addiction as well as other self-defeating behavior caused by the same failure to consider the big picture. It's a defense against the negativity bias—a rule that the rational brain can deploy against irrational impulses.

Like football coaches, we're too often swayed by a clear short-term gain rather than a nebulous long-term benefit. That's why procrastinators surf the web instead of working on the project due next month. That's why smokers and alcoholics choose immediate pleasure over their future health. The way out of any addiction, whether it's to safety or to a harmful habit, is to follow a rule instead of your gut.

When you're in a situation that doesn't lend itself to rules, you can take the decision out of your hands by following the guidance of someone else. In those loss-aversion experiments with gambling, when people were asked what bet someone else should make, they made smarter decisions than the bettor himself. The outsiders weren't fearful of losing their own money, so when the odds were favorable they were more willing to bet, and for higher stakes.

This was an important finding, because it showed that the negativity bias doesn't sway all decisions equally. The effect is stronger for things that matter more to us, like preserving and protecting the self and presumably also a romantic partner or a family member. But if the decision involves a stranger, as it did for the people in the gambling experiments, then the brain isn't as susceptible.

So when you're facing a potentially costly decision of any kind, you can avoid safety addiction by heeding someone who understands the risks but has nothing personal to lose, like a colleague or a friend or a counselor. In the stock market, for instance, one of the most common mistakes that investors make is to hold on to a falling stock for too long because they're reluctant to take the loss. This is a version of what economists call the sunk-cost fallacy, which occurs when individuals or companies refuse to give up on a doomed project because they don't want to write off all the time and money they've sunk into it. They'd be better off cutting their losses, but that decision is so painful that it often isn't made until an outsider comes in with a different perspective—and no personal stake in the loss.

Once Kelley overcame his safety addiction, once he followed a rule based on others' expert analysis, he discovered that the statisticians had actually underestimated the benefits. As predicted, his players usually succeeded on fourth and short, and they often succeeded on fourth and long, too, resulting in scoring drives that more than made up for the times they failed. But that wasn't all. One fringe benefit was in the way they drilled during the week: They didn't have to practice punts anymore, which gave them an extra twenty-five minutes every day to work on other plays. Another advantage was extra flexibility throughout the game. When a conventional offense faces third down and long yardage, the other team's defense can drop back, concentrating on stopping a pass rather than a run, because it knows the offense has only one chance left. Defending against Kelley's team is much tougher. It can do whatever it wants on third down—run or pass, short or long—because it knows it'll have another chance on fourth down.

The team also benefits from an unexpected psychological effect. Coaches love to talk about the "morale factor" when explaining why they don't heed statisticians. It has been one of their standard excuses for playing it safe on fourth down: If the team takes a risk and fails, the players will get discouraged and their performance will suffer. That sounds plausible—bad events, after all, do have a disproportionate emotional impact—but it didn't happen to Kelley's team. Its play didn't deteriorate after a failure. When Kelley analyzed the numbers to gauge the impact of his fourth-down strategy, the only psychological effect he found was on the *other* team.

If the opponent's defense failed to stop Pulaski on a fourth down, it suddenly became much more likely to make a mistake and yield a long gain on one of the next few plays. Kelley's players could handle a setback on fourth down because he'd taught them to look at the big picture, but the other team was flummoxed because it was focusing on the one failure. Kelley hadn't merely taught his players to overcome their negativity bias. He was using the power of bad to psych out the opposition.

The sum of these advantages has created a new powerhouse in Arkansas. When a conventional high-school offense gets the ball, it will go on to score a touchdown or a field goal about a third of the time. Kelley's will score three times out of four, averaging about fifty points a game, and those points virtually all come from touchdowns, because his strategy forbids kicks of any kind. Other coaches, when facing a fourth down near the other team's goal line, will settle for the safety of a three-point field goal rather than risk going for the six-point touchdown. But Kelley won't settle except in a situation when the odds overwhelmingly favor a field-goal attempt, which happens about once a season. The rest of the time, no matter where his team is on the field or how many yards it needs to gain, he goes for it—with phenomenal success.

During the three decades before Kelley's arrival, Pulaski Academy reached the semifinals of the Arkansas state championship only twice, and lost both times. In Kelley's sixteen seasons, the team has reached the semifinals thirteen times and gone on to win the state championship seven times. The record has earned him national recognition, including the All-USA Football Coach of the Year award from *USA Today,* and he gets speaking invitations around the country to explain his radical strategy.

Yet he still hasn't converted other coaches. In a rational world, his record would be enough to inspire imitators, but safety addiction is not rational. "Coaches will listen to me and tell me I'm right about the odds, and then they'll explain why it wouldn't work for their team or why they can't afford to risk losing their jobs," he said. "The loss aversion extends right up the ranks. I've discussed coaching jobs at the college level, at schools that you'd think would have nothing to lose—their teams never win anyway—but the athletic directors are afraid they'll be out of a job if they hire me and it doesn't work out."

The only converts he has made are at home, among the fans who used to boo him when the team failed on fourth down. During his first season, there was so much hostility to his strategy that on the rare occasion of a punt, the fans cheered wildly. But they've now seen the long-term results. His overall positivity ratio—192 wins versus 27 losses—is way above 4 to 1, and even die-hard traditionalists have been convinced. "If anyone starts booing now," Kelley said, "people in the stands will turn around and shout, 'What are you booing, idiot? We win all the time playing this way.'"

That's one strategy to subdue the power of bad: overwhelm it with good. There are various techniques for accentuating the positive, and we'll be discussing them later. But first we'll focus on a more efficient strategy: eliminating the negative. Because of that 4-to-1 ratio, you get a lot more leverage by starting with the bad stuff.

CHAPTER 2

Love Lessons

Eliminate the Negative

There's an elegant symmetry to the traditional wedding vows: for better or for worse. But love is not symmetrical, and most of us don't realize how lopsided it can be. The worse matters far more than the better in marriage or any other relationship. A slight conflict can have ruinous consequences when the power of bad overwhelms your judgment, provoking you to actions that further alienate your partner or your friend or your child. The negativity effect magnifies their faults, real or imagined, starting with their ingratitude, because you're also biased by that internal overconfidence that magnifies your own strengths. So you wonder how your partner or your friend or your child can be so selfish and so blind to your virtues—to all that you've done for them. You contemplate one of life's most exasperating mysteries: *Why don't they appreciate me?*

We have some answers, thanks to psychologists who have been tracking couples' happiness. They've found, based on the couples' ratings of their own satisfaction, that marriages usually don't improve. The ratings typically go downhill over time. The successful marriages are defined not by improvement but by avoiding decline. That doesn't mean marriage is a misery. The thrill of infatuation fades, so the euphoria that initially bonded a couple cannot sustain them over the decades, but most couples find other sources of contentment and remain satisfied overall (just not as satisfied as at the beginning). Sometimes, though, the decline in satisfaction is so steep that it dooms a marriage. By monitoring couples' interactions and tracking them over time, researchers have developed a surprising theory for the breakdown of relationships—or at least a theory that surprised the researchers and the bickering couples. It wouldn't have come as a shock to Anthony Trollope, who published his own version of it in 1869 without the benefit of any lab experiments.

It appeared in *He Knew He Was Right,* a novel that was too prescient for its own good. It failed to resonate with Victorian critics and readers because the story seemed so unlikely: a happy family destroyed for no reason. Louis and Emily Trevelyan have everything going for them. They're young and in love. Both are gorgeous, intelligent, honorable, and well-bred. Emily, the daughter of a diplomat in a remote colony, lacks social position and money, but Louis has both and is glad to share with the wife he adores. As the newlyweds settle in London, they seem the perfect couple except for a hint that Trollope drops during a conversation between Emily's parents, when the mother observes that their son-in-law "likes to have his own way."

"But his way is such a good way," the father says. "He will be such a good guide."

"But Emily likes her way too," the mother replies.

The first two years of the marriage bring a baby boy and continued bliss, until one day Emily receives a visit from a middle-aged bachelor who's an old friend of her father's. Emily thinks nothing of it, which is understandable because she's new to the London social scene. Louis is concerned, though, and not because he doubts his wife's fidelity. This bachelor has a reputation as a Lothario, one whose visits to another married woman's home created a scandal that forced the family to flee London. When Louis explains the situation to Emily, expecting her to recognize that any more visits would cause gossip, she doesn't take him seriously, insisting that no one could suspect her. Exasperated by her nonchalance, Louis loses his temper and declares he wants no more visits.

His outburst leaves Emily feeling hurt and humiliated. How could her husband treat her with such contempt? How, after all her devotion to him, could he impugn her honor? During a subsequent visit, the bachelor starts flirting, making Emily realize that Louis was right, but she still can't forgive her husband's rudeness. She tells herself that she'll end the visits only if Louis first apologizes for treating her so badly.

But now Louis is in no mood to apologize. How could his wife see this Lothario yet again? How could she be so heedless of his feelings? He asks the eternal question—*Why doesn't she appreciate me?*—and conducts an inventory of the relationship:

> And then he began to recapitulate all the good things he had done for his wife, and all the causes which he had given her for gratitude. Had he not taken her to his bosom, and bestowed upon her the half of all that he had simply for herself, asking for nothing more than her love? He had possessed money, position, a name,—all that makes life worth having. He had found her in a remote corner of the world, with no fortune, with no advantages of family or social standing . . . but he had given her his heart, and his hand, and his house.

The inventory is accurate, but ultimately Louis's calculations are just as faulty as Emily's when she tallies up all her acts of devotion. Their marriage is doomed because they're counting the wrong things.

Love on the Rocks

Imagine you are dating someone who does something that annoys you. (This may not require a great deal of imagination.) Perhaps your partner is a spendthrift, or flirts with your friends, or zones out in the middle of your stories. Or perhaps your partner is a tightwad who clams up around your friends but bores you with stories that never get to the point. How do you respond?

a. Let it slide and hope things improve.
b. Explain what bothers you and work out a compromise.
c. Sulk. Say nothing but emotionally withdraw from your partner.
d. Head for the exit. Threaten to break up, or start looking for another partner.

Those answers form a matrix used in a classic study of how dating couples deal with problems. Psychologists at the University of Kentucky identified two general strategies, constructive or destructive, each of which could be either passive or active. The constructive strategies sounded sensible and admirable, but they didn't matter much. Remaining passively loyal had no discernible impact on the course of the relationship; actively trying to work out a solution improved things only a little.

What mattered was the bad stuff, as the psychologists concluded: "It is not so much the good, constructive things that partners do or do not do for one another that determines whether a relationship 'works' as it is the destructive things that they do or not do in reaction to problems." When you quietly hang in there for your partner, your loyalty often isn't even noticed. But when you silently withdraw from your partner or issue angry threats, you can start a disastrous spiral of retaliation.

"The reason long-term relationships are so difficult," said Caryl Rusbult, who led the couples study, "is that sooner or later one person is liable to be negative for so long that the other one starts to respond negatively too. When that happens, it's hard to save the relationship." Negativity is a tough disease to shake—and it's highly contagious. Other researchers have found that when partners are separately asked to ponder aspects of their relationship, they spend much more time contemplating the bad than the good. To get through the bad stuff, you need to stop the negative spiral before it begins.

But suppose you've managed to survive your courtship without any problems. (This may take more imagination.) You've just graduated from dating to blissful matrimony. Your soul soars, your heart sings, and your brain is awash in oxytocin, dopamine, and other neurochemicals associated with love. You are probably in no mood to participate in a scientific study, but some other newlyweds were persuaded to do so for a long-term project called PAIR. (The full unromantic name was Processes of Adaptation in Intimate Relationships.) These couples in central Pennsylvania were interviewed during the first two years of marriage by psychologists who cataloged both the positive and negative aspects of the relationships.

Some of the people were already ambivalent or hostile toward their partners—and tended to get divorced quickly—but most couples showed lots of mutual affection and went on to celebrate several anniversaries. Over the long haul, though, those tender early feelings were not a reliable harbinger. More than a decade later, a disproportionate number of the couples who had been "almost giddily affectionate" were no longer together. As a group, those who divorced had been a third more affectionate during the early years than the ones who went on to have long happy marriages. Over the short term, their passion had enabled them to surmount their misgivings and their fights, but those positive feelings couldn't keep the marriage going forever. Once again, it was how they dealt with the negative stuff—their doubts, their frustrations, their problems—that predicted whether or not the marriage would survive. Negativity hits young people especially hard, which is one reason that people who marry earlier in life are more likely to divorce than ones who delay marriage. (Another reason is that younger people tend to have less money, which means more stress.)

Some couples, of course, are better off splitting up, but far too many of them sabotage a relationship that could have worked. Researchers who track couples have repeatedly been puzzled to see relationships destroyed even when there are no obvious causes. To test a theory, the psychologists Sandra Murray and John Holmes brought couples into the lab and gave them questionnaires to be filled out at tables arranged so that the partners sat with their backs to each other. They'd both be answering the same questions, the experimenter explained, and it was important that they not communicate in any way as they filled out the forms.

In fact, though, the questionnaires were different. One form asked people what they didn't like about their partners. They could list as many traits as they wanted but were told it was fine to name just one. These people, who'd been dating on average for a year and a half, had a few complaints but were mostly pretty satisfied. They typically wrote down one or two things about their partners that were less than ideal and then they put down their pens. The other partner was given a much different task: listing all the things in their home. Instructed to name at least twenty-five items, they'd start writing—cataloging pieces of furniture, kitchenware, gadgets, books, artwork, whatever—and were often still working away at it when the experimenter returned five minutes later.

Meanwhile, the other partner was sitting there with nothing to do but listen to the scribbling—and assume that it must be a vast inventory of their personal failings. They'd been hard-pressed to name just one or two complaints, but their partner apparently had a much different view of the relationship. As always in such studies, the partners were later informed of the deception, so nobody went home unhappy. But before revealing the truth, the experimenters asked more questions about the relationship, and it turned out that that the deception had a big impact on some of the people: the ones already prone to insecurity. The people with high self-esteem (as measured in a test before the experiment) felt a little threatened but shrugged it off because they were secure enough to know that their partners valued them. But the people with low self-esteem reacted strongly to the presumed cascade of criticisms.

Once they heard all that scribbling behind their backs, they feared their partners might reject them, and that fear took over. To protect themselves, they changed their own attitudes. They lowered their regard and affection for their partner. They felt less close, less trustful, and less optimistic about the relationship. The insecure people were reacting needlessly, because in reality they were valued by their partners just as much as the secure people were. But they projected their own self-doubts into their partners' minds. They assumed their partners would judge them as harshly as they judged themselves.

This sort of needless self-protection is especially harmful to a relationship, as Murray and Holmes found in another study by tracking a group of newlywed couples over three years. All too often, couples would seem to be in good shape—they had relatively few conflicts—but then one partner's insecurities would kick in. They'd mentally push their partners away or devalue their relationships even though there was no real danger. They'd become especially resentful at making routine sacrifices, like staying home in the evening instead of going out with friends. Their relationships were among the strongest to begin with, but they fell apart rapidly, just like the marriage in *He Knew He Was Right*.

In Trollope's novel, no one else understands why Louis and Emily are so angry. Again and again, friends and family beg Louis and Emily to forget their grievances. But each attempt at reconciliation fails, like the letter Louis sends to Emily pleading to save the marriage. "I do not in the least suspect you of having as yet done anything wrong," he writes to her, "or having even said anything injurious either to my position as your husband, or to your position as my wife."

He expects her to appreciate this reassurance, but instead she focuses on just two words: He doesn't suspect her *as yet* of any sin. *As yet!* So he must suspect she's capable of adultery in the future! In her mind, those two words outweigh all the declarations of love in the rest of the letter, but Louis can't fathom why she remains upset, so he angrily proceeds to escalate the conflict. The novel's plot has left many readers as frustrated as the friends of Louis and Emily—how could two people in love keep misinterpreting and overreacting to each other?—but it's a remarkably accurate portrayal of how the negativity effect corrodes a relationship. "The truth was that each desired that the other should acknowledge a fault," Trollope writes. "When they met, each was so sore that no approach to terms was made by them."

By watching sore spouses bicker, researchers have noticed the same pattern of gender differences depicted by Trollope. Like Louis, insecure men tend to focus on fears of their partner's sexual infidelity. Inflamed with jealousy even when there's no cause for it, they become highly possessive and controlling, which puts stress on the relationship and eventually drives the woman away. Insecure women worry less about sexual infidelity than about other kinds of rejection, and they tend to react with hostility rather than jealousy. These various reactions were cataloged in a study of New York City couples who were videotaped in a lab at Columbia University as they discussed their problems.

Each time one of the partners did something negative—such as complaining, speaking in a hostile tone, rolling their eyes, denying responsibility, insulting the other—the action was classified and counted. The researchers, led by Geraldine Downey, found that insecure people were the ones most likely to act negatively. Their own fear of rejection no doubt intensified the distress they felt, because for them an argument wasn't just about a specific issue but a sign of deep problems and an ominous signal that the relationship was in jeopardy. They were so sensitive to the negativity bias that they were living in a state of anxiety. Their panicky response was to push away their partner—with unfortunate success, as the researchers found by following couples over several years. It turned out that people sensitive to rejection were especially likely to end up alone. Their fear of rejection became a self-fulfilling prophecy.

The negativity effect was confirmed in an even more elaborate study of couples in Seattle who were wired to record their pulse and other physical reactions, and filmed as they conversed about the events of the day and their problems. Afterward, the spouses separately watched the recordings and rated the positive and negative emotions they had felt at each point in the conversation. Three years later the researchers, Robert Levenson and John Gottman, tracked down the couples to see who was still happily married. It turned out, again, that negative moments were much more important than positive ones. The couples most likely to become unhappy were the ones who showed the highest levels of stress during their conversations, as measured either by physical arousal or by the level of negative emotions. Reciprocating positive feelings had little or no effect on the marriage's prospects, but how each spouse responded to the other's negativity was crucial. When a wife returned her husband's anger, or when a husband reacted to his wife's criticism by shutting down emotionally, the couple's relationship was likely to deteriorate.

Negativity seems to be less of a problem in same-sex couples. When the Seattle researchers tracked a group of same-sex couples for more than a decade, they found that both male and female couples tended to be more upbeat than heterosexual couples when dealing with conflict. They were more positive both in the way that they introduced a disagreement and in the way that they responded to criticism, and they remained more positive afterward. In heterosexual couples, the most common conflict pattern is called "female-demand-male-withdrawal," a destructive cycle in which the woman initiates a complaint or criticism and the man responds by withdrawing. That pattern is less likely in same-sex couples. If it's two men, they're less likely to initiate a complaint; if it's two women, they're less likely to withdraw after being criticized.

Whatever their orientation, many couples count on nonverbal ways to reconcile after fights, but researchers have found that even good sex can't be counted on to make up for bad arguments. A study of verbal and nonverbal communication between newlyweds showed that the amount of negative talk—anger, hurt, resentment—was a powerful predictor of marital happiness, whereas expressing affection verbally helped only a little, and expressing it sexually had no discernible effect. Other studies have found modest benefits to sex, although yet again the power of bad prevails. The psychologist Barry McCarthy, who is a couples therapist as well as an academic researcher, estimates that when sex is good, it can contribute 15 to 20 percent to the happiness of a marriage, whereas when sex is bad or nonexistent, it can become a much bigger factor, decreasing the couple's happiness by 50 to 75 percent. His conclusion jibes with the advice in *Cat on a Hot Tin Roof* given by the family matriarch, Big Mama. "When a marriage goes on the rocks, the rocks are there," she says to her daughter-in-law, pointing to the bed. "Right *there*!"

Most people don't recognize the negativity effect in their relationships. When Baumeister asks his students why they think they would be a good partner, they list positive things: being friendly, understanding, good in bed, loyal, smart, funny. These things do make a difference, but what's crucial is avoiding the negative. Being able to hold your tongue rather than say something nasty or spiteful will do much more for your relationship than a good word or deed.

The Good-Enough Relationship

There's a comforting lesson in the marital research, and it applies not just to marriages. Social scientists have found that the negativity bias governs lots of other relationships: parent-child, teacher-student, mentor-protégé, boss-employee. It governs casual relationships, too, like the ones with your neighbors. One of the earliest findings in the field of social psychology was the "propinquity effect," which emerged from a study in the 1940s that meticulously mapped the formation of friendships among neighbors in a newly built community of houses and apartments in the Boston area.

It turned out that friendships depended on the same factors as real-estate prices: location, location, location. People were most likely to become friendly with their next-door neighbors, and the odds of a friendship decreased sharply with distance, so that people living just a few doors apart rarely socialized. The propinquity effect (also known as the proximity principle) seemed a cheery addition to Psych 101 textbooks: solid evidence of how genial people can be. Let their paths cross often enough, and they'll find common ground.

Eventually, though, some other researchers took a look at another community, a town-house development in a Los Angeles suburb, and studied feuds as well as friendships. It turned out that people's enemies lived even closer than their friends. There was a dark side to the propinquity effect. When it came to getting along with the neighbors, bad was stronger and quicker than good. By tallying up social interactions, the researchers found that it took a lot of socializing to build a friendship, but one or two bad encounters could turn neighbors into permanent enemies.

That finding wasn't such a cheery addition to psychology textbooks, but there is a practical and reassuring lesson to be drawn from it and the other discoveries about the negativity bias in relationships. Whether you're dealing with a neighbor, a spouse, a child, or anyone else, the lesson is: You don't have to try so hard.

You don't have to pay so much attention to the therapists who have been ordering everyone to work on their relationships. Husbands and wives are assigned task after task: Have sex every day, make time for romantic dinners and exotic vacations, take long walks on the beach and bare your souls. Parents have still more chores. Ever since Freudians started blaming mothers and fathers for their children's neuroses, parents have felt enormous responsibility to do everything right. But the recent research into the negativity effect offers relief, because it shows that these efforts are overrated.

For overworked parents, some of the most encouraging evidence comes from research into intelligence, which is probably the best understood and most precisely measured psychological trait. Psychologists have looked at various populations—like identical twins growing up separately, or children raised by adoptive parents—to analyze the two major factors that determine a child's IQ: genes and environment. After gauging children's genetic potential (based on the IQs of the biological parents), researchers can see how the children's cognitive development is affected by the homes in which they grow up. Affluent and well-educated parents can provide their children with personal instruction, lots of books, and access to good schools and enrichment programs. If the parents lack education and are struggling to make ends meet, the children have access to fewer intellectual resources and must contend with more obstacles.

Studies have repeatedly shown that when children grow up in low-income households headed by adults who are not well educated or professionally accomplished, then the children are less likely to reach their full intellectual capacity. Some of them thrive despite—or even because of—the challenges, but on average their IQs are lower than would be predicted from their genetic potential. In fact, some studies have found virtually no correlation between their genetic potential and their actual IQ, because in so many cases the negative environment overwhelms the positive genes.

But the environment barely matters among parents higher on the socioeconomic scale. Their children's IQs are determined almost entirely by genes. It doesn't much matter whether the parents have bachelor's degrees or doctorates, whether they're middle class or rich, whether they're merely competent in their careers or highly successful. Above a certain level, the parenting doesn't have much impact on the child's intellectual capacity. Well-off parents may serenade the womb with Mozart, hire the best tutors, and pay for the most expensive schools, but they can't significantly raise their children's IQs. By avoiding a negative environment, they can make sure their children's genetic potential is fulfilled, but that's about it.

The same pattern holds for the emotional aspects of child rearing. The Yale psychologist Sandra Scarr and other researchers have found that as long as parents avoid being violent, abusive, or neglectful, then it doesn't matter significantly what else they do. Bad parenting scars children, but being especially conscientious doesn't reliably make children happier or healthier. Nor does it make much difference during adolescence, as a team of Dutch researchers concluded after interviewing a large sample of teenagers about their parents' good and bad qualities. The good qualities, like being loving and emotionally supportive, had little effect. What mattered, at least when it came to predicting which kids were unhappy or got in trouble with the law, were bad qualities like getting angry and disciplining too harshly and unfairly.

Because of these findings, a few psychologists have been advising parents to give themselves a break. Instead of trying to be perfect, instead of feeling obliged to attend every soccer game and help with every school project, just be a "good enough" parent. Your child will still do fine, and you'll reap benefits in your marriage and the rest of your life. The good-enough concept hasn't caught on with today's legions of helicopter parents and tiger mothers, but it's excellent advice that deserves to be extended to other relationships, too.

Be a good-enough husband, a good-enough wife, a good-enough friend or neighbor, a good-enough teacher or boss. Focus not on achieving perfection but rather on avoiding elementary mistakes, both in your behavior and in the way you interpret others' behavior. Here are some specific strategies:

Don't overpromise. Most of us tend to promise more than we can deliver because of what psychologists call the planning fallacy, which is our tendency to underestimate how much time and effort will be required for a task. When we don't deliver on time, we hope that our family or friends or colleagues will at least appreciate our good intentions—*See how much I was trying to do for you!* But they won't. The negativity effect is in force. They'll focus not on your good intentions but on the bad result.

Don't expect credit for going the extra mile. When you do more than you promised, your generosity is likely to go unappreciated, as Ayelet Gneezy demonstrated in experiments inspired by Amazon.com. She'd noticed that when her packages from Amazon arrived earlier than promised, she didn't feel particularly grateful. She and Nicholas Epley, a colleague at the University of

Chicago, found that this was a common reaction in their experiments with students.

The students would blame a ticket broker for selling them worse seats than promised, but they wouldn't show extra appreciation if the seats were better than promised. In one experiment, when students were paid for each puzzle they solved, another person would volunteer to help (for no charge) by solving ten of the puzzles. If he fell short, they would fault him for being untrustworthy and not making enough effort, but he would get the same ratings whether he fulfilled his promise or went beyond it and did an extra five puzzles. Doing the extra puzzles clearly required more work, but the recipients of his generosity didn't appreciate it.

Remember that bad is in the eye of the beholder. An offense that seems trifling to the rest of the world can destroy a relationship if it looms large to one person. You have to deal with your partner's reaction even when it makes no sense to you. As we've seen in the couples research, marriages often fail because men and women tend to focus on different kinds of bad. Men worry more often about their partners' sexual infidelity, while women worry more often about their partners' emotional withdrawal. Other gender variances have emerged in the large research literature into personality.

Studies in dozens of countries around the world have shown that the biggest personality differences between men and women involve negative emotions. Women experience anger, anxiety, and depression more frequently than men do. They're also better than men at detecting others' negative feelings. When people are shown faces registering different emotions, men and women are equally adept at identifying happy faces, but women are significantly more accurate in spotting fearful, sad, and angry faces. There's an evolutionary explanation for this female advantage: The survival of infants has long depended on the ability of their mothers to recognize nonverbal signals of distress. (Also, since women are generally smaller than men, there could have been an evolutionary advantage in being able to spot another person's anger earlier on, before it reaches the point of physical violence.) A husband who withdraws emotionally may think he's buying marital harmony by hiding his negative feelings, but his wife is liable to see through him.

This emotional disparity often leaves men unable to understand what's gone wrong. That's the great failing embodied in Trollope's title, *He Knew He Was Right.* Louis is indeed right about the facts—he's quite correct in warning his wife not to endanger her reputation—but he goes about it all wrong because he doesn't consider how his angry admonition will affect her. "He should have spoken to her gently," Trollope writes, "and have explained to her, with his arm round her waist, that it would be better for both of them that this friend's friendship should be limited. There is so much in a turn of the eye and in the tone given to a word when such things have to be said,—so much more of importance than in the words themselves." Louis is correct in his own eyes, but the marriage is in trouble because his wife sees it so differently—and Louis is too obtuse to recognize the pain he's causing her. During an argument, it's better to study your partner's reaction and imagine their perspective than to keep repeating your grievance. Talk less. Listen more.

Put the bad moments to good use. Our brains evolved to focus on the negative because it's the most effective way to learn. When something goes wrong in a relationship, don't despair that you're not meant for each other. Look for a lesson. Take criticism not as a malicious attack but as useful feedback. When Louis warns

Emily that she's endangering her reputation by seeing her father's friend, he's giving her valuable information. As a newcomer to London society, she needs to learn its ways and avoid its gossip. By treating his warning as an indictment of her character and a sign the marriage is in trouble, she's passing up a lesson that could help her personally and strengthen their marriage. Louis could likewise have gained something from their first fight if he'd seen how his peremptory tone needlessly insulted her, but he, too, passed up the chance to learn a lesson that could have averted the next fight. Bad can be a great teacher.

Think before you blame. Beware the *fundamental attribution error,* which is psychologists' term for an all-too-common mistake. Suppose a couple has planned to meet for a romantic dinner at a special restaurant and one of them shows up late. The tardy one will blame it on a specific situation, like a crisis at the office or an unexpected traffic jam. But if you're the one waiting alone in the restaurant, you're liable to read more into it. "Just *typical,*" you tell yourself, fuming at this demonstration of your partner's general unreliability, selfishness, or worse ("She doesn't love me anymore").

Maybe you're right, or maybe you're making the fundamental attribution error, which is the tendency to draw conclusions about someone else's inner character based on behavior that's actually due to an external situation. When we run a stop sign, we factor in the mitigating circumstances ("A tree blocked my view of the sign"), but when we see someone else do it, we hold them fully responsible ("What a lousy driver!").

Researchers originally theorized that the error applies to all kinds of behavior, both good and bad, but eventually they recognized that familiar negativity effect: The error occurs only when drawing conclusions from negative behavior. In fact, when someone else does something positive, we're apt to read less into it ("He got straight As because his professors are easy graders"), whereas we give ourselves full credit ("I got As because I'm a good student"). It's when we're judging someone's negative actions that we tend to draw unfairly broad conclusions, even when it's someone that we know intimately.

So instead of enjoying that romantic dinner, you sulk at the rudeness of your tardy partner, who feels unjustly blamed for something beyond their control. Over the long haul, this sort of mutual misunderstanding is deadly to relationships, as the psychologists Benjamin Karney and Thomas Bradbury found when they classified newlyweds according to their "attributional style." The ones most prone to attributing their partners' slipups to permanent internal characteristics rather than temporary external factors were much more likely to wind up divorcing.

It's not easy overcoming this attribution error—that's one reason it was dubbed fundamental—but once you're aware of it, you can consciously compensate. Before automatically blaming your partner's mistake on a character flaw, or interpreting it as a symptom of a permanent problem, force yourself to consider more benign explanations. And then give your partner the benefit of the doubt.

When you're fighting, bring in an imaginary referee. Another way to shift your thinking is with the "marriage hack," the term Eli Finkel uses for a simple technique that his team of social psychologists taught to married couples in the Chicago area. Every four months, the couples were asked about their satisfaction with the marriage and their biggest recent disagreement. Over the first year of the study, their marital satisfaction slowly declined, as it typically does. Then, at the start of the second year, when the couples reported on their latest conflict, some of them were given an additional task: Imagine you're a neutral third party observing this disagreement. How would you see it? Could you find any good that might come of it?

After writing up this imaginary referee's verdict, those couples were instructed to try taking this third-party perspective during their arguments at home. It worked. Their marital satisfaction stopped declining and remained stable over the next year, in sharp contrast to the continuing decline among the other couples in the study. By taking the imaginary referee's perspective, the couples tempered their anger and became more willing to listen and compromise.

Get a second opinion. If the imaginary referee doesn't work, try a real one. Shifting perspective can be especially difficult when your own insecurities are skewing your vision, but you can always get a second opinion from someone whose judgment isn't warped by personal fears of rejection. That's one of the chief benefits of couples therapy, but you don't need to hire a professional to get another opinion. In Trollope's novel, the whole crisis could have been avoided if Louis or Emily had heeded their friends. The single most valuable piece of advice in the novel comes free of charge from Emily's sister, who realizes that Louis didn't mean to insult Emily and tells her: "If I were you, I would forget it."

Suspend judgment. Being able to overlook your partner's offenses, imagined or real, is one of the surest ways to stay married. A friend of ours keeps his wife's faults in perspective by taping a message to his bathroom mirror: "You're no bargain either." Others seem to do it automatically, as researchers were surprised to discover by scanning the brains of lovers. The social psychologists Xiaomeng Xu and Arthur Aron teamed with the neuroscientist Lucy Brown to study couples in Beijing who were still in the early stage of infatuation. When these people were shown a photo of their partner as well as a photo of someone else, the image of their beloved triggered lots more activity in the circuits associated with the dopamine-reward system, the same system activated by cocaine and other intense pleasures.

But what happened once the infatuation wore off? Three years later, only half of the couples were still together. To understand their longevity, the researchers went back to these lovers' original data and compared it with that of the couples who would later break up. A crucial—and unexpected—difference emerged: When the people destined for romantic success looked at a photo of their partner, there was reduced activity in a region of the prefrontal cortex used in making negative judgments. Their critical faculties were suspended, and the relationship thrived.

This finding was confirmed in a brain-scanning study of couples in New York City who had been married for over two decades. The social psychologist Bianca Acevedo and the anthropologist Helen Fisher found that the spouses who were most satisfied with their marriage (as measured on a questionnaire) showed less activity in this same brain region associated with negatively judging other people. They, too, tamped down their negativity bias when looking at the photo of their beloved. Their spouses undoubtedly had flaws, but their marriages happily endured as their brains automatically followed a principle described by the poet William Blake: "Love to faults is always blind."

Blake's wisdom has been further confirmed in other studies tracking the illusions of couples. These illusions were measured by asking people to rate their own qualities as well as their partners' qualities. The most unrealistic couples—the ones who idealized their partners, giving them higher ratings than the partners gave themselves—were most likely to remain satisfied and stick together over the course of the study. And there was an added bonus from this positive illusion: Eventually the partners would come to share this idealized view of themselves. Everyone felt better.

If these positive illusions don't come naturally to you, cultivate them by consciously reminding yourself of your partner's strengths. When she demonstrates one of those strengths, tell her you admire that quality. When you're irritated by one of his habits, try to understand why he does it, and look for an upside: Sure, he obsessively overpacks for vacations, but at least he doesn't forget anything, and you may end up needing some of that stuff yourself.

Don't take the bait. Even if your brain's circuits detect your partner's sins, you can still act as if you didn't notice. You can follow the advice that Ruth Bader Ginsburg received on her wedding day from her mother-in-law: "In every good marriage it helps sometimes to be a little deaf." More than half a century later, Ginsburg passed that advice on to an audience of college students, explaining that she had followed it assiduously not only at home but in every place she worked, including the Supreme Court.

"When a thoughtless or unkind word is spoken, best tune out," Ginsburg told the students. "Reacting in anger or annoyance will not advance one's ability to persuade." You can't tune out all the time, of course, because some affronts are too painful to endure. Some problems need to be confronted, but a quick angry reaction isn't the way to do it—especially not in the age of Twitter.

If you must respond, don't escalate. Stand up for yourself and explain what's bothering you, but stay calm. Again, remember that what seems terrible to you can seem harmless to everyone else, so don't accuse the other person of rudeness or malice. (Later we'll discuss how to deliver criticism and bad news.) Don't sulk and don't retaliate, because a small disagreement can quickly turn into a major fight. Because of the negativity effect, there's a natural tendency for conflicts to escalate, as the psychologist Boaz Keysar and colleagues at the University of Chicago neatly demonstrated by testing two versions of a game called Dictator.

In the standard version of this game, one player is given $100 and told he can share his windfall with the other player. He's called the dictator because he can do whatever he wants—split it with the other player, give her a smaller share, keep it all himself. After he decides, they switch roles and play another round in which she gets a new $100 windfall for her to divide between them. Then they play more rounds, taking turns being the dictator.

Things went smoothly in the Chicago experiment when this standard version was played. The dictator would typically start off by keeping a bigger share of the money for himself, but he gave enough away to satisfy the other player. The next round, she typically reciprocated by giving him at least as much as he'd given her and then he would respond with similar generosity. As they played more rounds, the players moved closer to a 50-50 split.

But things weren't so friendly when the researchers tried a negative version of the game in which the windfall was described differently. Instead of calling it a gift to the dictator, the researchers said that the $100 had been given to the other player, and that the dictator would now decide how much of it she got to keep. It was still the same decision—the dictator was still dividing $100 between them—but it was presented as a loss rather than a gain. Instead of receiving a gift from the dictator, the other player was having her own money taken away.

In this negative version of the game, the dictator typically started off being more generous than in the standard version. He realized that the other player would resent having her money taken away, so instead of taking a bigger share himself, he typically split it 50-50. But that still didn't sit well with her. The next round she typically kept more, splitting it about 60-40, and things deteriorated in succeeding rounds as the players got more and more selfish. They didn't merely retaliate against each other for a perceived affront. They escalated. The results of the experiment inspired the Chicago psychologists to suggest a revised rule for reciprocity: *You scratch my back and I'll scratch yours, but if you take my eye, I'll take both of yours.*

So before you respond, control your anger. Instead of retaliating, you might say something like, "I'm going to try to stay positive, but I can't keep doing that unless you do, too." Whatever you say, remember that as bad as things seem at the moment, either of you can easily make them a lot worse.

Follow the Negative Golden Rule. Most of us like to think well of ourselves, and we assume our sterling qualities are obvious to our friends and family, too. We expect them to pay attention to the favors we do, the sacrifices we make, the joys and the laughter we bring to their lives. But that's not what seems most obvious to them.

Avoiding bad is far more important than doing good when you're dealing with lovers, children, friends, colleagues, or anyone else. It's not so much what you do unto others. It's what you *don't* do.

CHAPTER 3

The Brain's Inner Demon

Wired for Bad

Before Felix Baumgartner aspired to leap off a balloon capsule twenty-four miles above Earth, he was already known to his fans as Fearless Felix. He had jumped off the two highest buildings in the world and skydived across the English Channel wearing a carbon-fiber wing. He had stood at the rim of a six-hundred-foot-deep cave in Croatia and plummeted into the black void, emerging as fearlessly triumphant as ever.

The prospect of the first supersonic leap from the stratosphere didn't faze him either, not at first. He eagerly went to the Mojave Desert in California, near the air force base where the first supersonic flights took place, to train under the guidance of aerospace experts. When asked about the risks, he calmly ticked off the ways he could die during the mission. The enormous gossamer balloon, forty acres of plastic thinner than a dry-cleaning bag, might be destroyed by surface winds in the first several thousand feet of the ascent, sending him crashing to the ground with no chance to deploy a parachute. If his suit lost pressure in the thin air at 120,000 feet, his blood would start to boil, but even that wasn't as much of a worry as the possibility of losing control of his body during the fall. As he accelerated to more than seven hundred miles per hour, part of his body would break the sound barrier while the rest of it was still subsonic, possibly creating turbulence that would cause a fatal "red-out," in which the body spins so fast that blood starts spurting out of the eyes.

"We just don't know what to expect when the body goes supersonic," Felix said one afternoon during his training. "I do have fears. But I have learned to control my fear so that it doesn't get in the way." So he thought, anyway. At age forty-one, he'd made more than 2,500 jumps as an Austrian military paratrooper and professional daredevil, but for this one he had to deal with a new challenge: a customized space suit and helmet. When he wore it during training sessions on the ground, he felt isolated because he couldn't feel the outside air or hear anything except the sound of his own breathing. At moments he felt trapped and desperate to get out of it, but he remained confident nonetheless.

"It's a mind thing," he said. "I'm not worried because I'm really good in preprogramming my mind and putting myself through fire and going the extra mile."

His confidence seemed justified later that day when he went into a wind tunnel for his first flight test. The bulky pressurized suit and helmet made any movement awkward—waddling into the wind tunnel, he looked like an arthritic version of the Incredible Hulk—but once the air lifted him off the ground, he expertly maneuvered his body into the proper falling position, headfirst at a forty-five-degree angle. Afterward, he and the engineering team were exultant. It seemed clear to everyone that he had the classic astronaut Right Stuff. They were good to go for the next step, a simulation of the five-hour ascent into the thin, frigid air of the stratosphere by putting Felix and his pressurized capsule in a high-altitude chamber at an air force base in Texas.

On the eve of the trip to Texas, Felix packed a suitcase at his apartment in Los Angeles, but that night he couldn't sleep. He kept thinking: *Five hours.* Could he make it that long locked inside the suit and helmet? By dawn he had his answer: *No way.* At six A.M. he booked a flight home to Austria and went to the Los Angeles airport. From there he telephoned Art Thompson, the aerospace engineer directing the project.

"Art, I'm sorry, I can't do this," he said. At first Thompson thought it was a joke. Then he tried to calm him down.

"Don't move," Thompson told him. "I'll go to the airport."

"No," Felix said. "I can't do this."

By this time he was sitting on the floor of the airport and crying. His sponsor, the Red Bull energy drink company, had spent untold millions, employing dozens of engineers, physiologists, and technicians for three years on the project, and he was failing them all. His daredevil career was over.

Fearless Felix had finally succumbed to the power of bad. He wasn't afraid of someday going into space and jumping twenty-four miles, but today he was too terrified to wear a space suit for five hours sitting safely on the ground. He'd managed to hide his fear during the years of training, but all the while it had been growing. At first it was just a mild discomfort a few minutes after he put on the suit and helmet. Then he began dreading the helmet ahead of time. As he drove from his apartment in Los Angeles, he'd tense up when he saw the Getty Museum and knew that he was only an hour away from the training site in the desert. He'd feel another jolt of angst when he reached the outskirts of Lancaster and saw Lake Palmdale, which meant he was just fifteen minutes away. He began thinking of the lake as the entrance to death row.

By the time he went inside the building to put on the suit, the tiniest things got to him. The neon light in the dressing room was too bright. He couldn't stand the smell of the rubber in the helmet's seal. He managed to convince everyone he had the Right Stuff, but eventually he couldn't keep pretending. The shame of fleeing was preferable to being imprisoned in that suit.

Why did the Wrong Stuff prevail? What happened to the self-confidence he'd professed early in his training with the suit? Felix couldn't say exactly, but he'd gotten one thing right in that earlier conversation: *It's a mind thing.*

The Fearful Brain

The brain evolved to protect the body. It was built rather slowly, mostly by adding on structures rather than replacing older ones, enabling it to create new systems for dealing with new dangers. Neuroscientists have identified three major threat-warning systems. The simplest and in evolutionary terms the oldest is the basal ganglia system. These clusters of neurons are already up and functioning in simple reptiles. They probably operate in a fairly automatic, hard-wired manner. They just detect standard sorts of threats and sound the alarm, which is enough to get the animal to fight or flee. Crocodiles don't respond to threats by holding committee meetings, drawing up contingency plans, or negotiating with the other side.

Later in evolution came another threat-alarm system often called the limbic system. It includes the amygdala, a small but influential structure in the middle section of the brain. It triggers emotions in response to threats, dangers, and problems. It doesn't replace the basal ganglia, but it operates alongside it, adding new levels of understanding and flexibility to the creature's responses.

These systems, which evolved long before language, respond even when we're not consciously aware of a threat. Neuroscientists have found that the sight of a hostile face can trigger a reaction in the amygdala before it registers in the rest of the brain. The basal ganglia probably operate in a pre-emotional manner and evolved to overreact. They often cause the animal to flee when it would have been safe to stay still, but that kind of mistake is far less costly than failing to escape from a predator. Suburbanites today don't face a lot of deadly threats around the house, but the ancient parts of the brain still react the same way: better safe than sorry.

Much later in evolution, a third alarm system emerged in the prefrontal cortex. The "pre" refers to its position in the brain, in front of the front, and thus presumably something quite new in evolutionary terms. It's far more developed in humans than in other species. The prefrontal cortex is involved in all sorts of very human, nonreptile activities, such as logical reasoning, conscious thought, and executive control. It can assess the symbolic meaning of events, including thinking with language. It is thus much more flexible and specialized than the basal ganglia or the limbic system, but it can activate those older systems. If a thought—*That lake means I'm only fifteen minutes from the suit*—triggers a basic sense of fear or danger, the person may react emotionally and irrationally. The prefrontal cortex can also process happier thoughts—*That suit will help me set a world record*—but the results won't be nearly so strong or quick, because bad rules throughout the brain and the nervous system.

When you look at the world, your attention is automatically drawn to threats. At the age of just eight months, infants will turn more quickly to look at an image of a snake rather than a frog, and at a sad face rather than a happy face. By age five, when shown a group of faces, they're quicker to spot a sad face than a happy face, and even quicker to spot a fearful or angry face. When adults put on goggles that show them a different image in each eye—an experimental method called binocular rivalry—the brain focuses longer on an image that evokes fear or disgust. In one goggle experiment, researchers used a photograph of a student after first passing on some gossip about him. Sometimes they'd mention that he had helped an elderly woman with her groceries; other times they'd say he had thrown a chair at a classmate. Then they'd show the student's photograph to one eye and a picture of a house to the other eye, measuring how long the brain focused on each image as it switched back and forth. When the student had been described as a chair-thrower, the brain focused on his face significantly longer than if he was a Good Samaritan. The brain was literally blinded to goodness.

This neural bias also shows up in experiments with the Stroop test, a classic technique for studying the brain's automatic processes by showing people words and asking them to identify the color of the letters. The trick to answering quickly is to ignore the meaning of the word and focus only on the color, but that's tougher to do with some words. If you see *green* spelled in red letters, you have to override your brain's automatic interpretation of the word's meaning, so it takes you longer to answer *red.* The same sort of delay occurs when people are shown words for negative traits. It takes them longer to answer for a word like *dishonest* than for a word like *friendly,* because the brain automatically pays more attention to bad traits than good traits. That bias affects their memories, too. When they're asked later to recall as many of the words as they can, they're twice as likely to remember a word like *dishonest* than *friendly.*

The negativity bias extends even to the simple neural signals traveling through the spine that control the reflexes to withdraw, as when your hand touches something sharp or hot. These are primal reflexes in humans as well as animals that evolved along with extensor reflexes that cause the creature to move forward. But withdrawal reflexes are much stronger physically than extensor reflexes and brook less interference. There's a similar imbalance in the amygdala, that ancient emotional part of the brain involved in basic processes like wanting things. (In the Freudian era, one would have associated it with the id, the well of desire and impulse.) The amygdala is important for making evaluations, because whether something is seen as good or bad depends on what your wants and needs are. When you're hungry, food looks and tastes good because it satisfies your desires. The amygdala is involved in unconscious processing but can also be affected by conscious instructions and goals, so its responses can be regulated to some degree.

To test its flexibility, researchers measured activity in the amygdala of subjects who were shown photographs of famous people. When the subjects were told to look for positive images, the amygdala was more active when seeing a beloved movie star or Mother Teresa than when seeing a villain. Conversely, when the subjects were told to look for negative images, the amygdala responded more strongly to pictures like Adolf Hitler's. So the amygdala's emotional reactions could be moderated by the conscious mind (and the experimenter's instructions). But the power of bad was still obvious. When the person was looking for bad things, processing of positive information was reduced, suggesting that the brain is pretty skilled at suppressing good news when it's irrelevant. But the brain is not so good at suppressing irrelevant bad news. It responded to good people only when looking for goodness, but it never stopped responding to bad people.

Even when things are going your way, the amygdala keeps looking for the cloud behind the silver lining. Experimenters have found that when subjects make a series of choices that either win or lose them money, the amygdala remains on alert for potential losses no matter how much money has been made. Even after someone makes all the right profitable choices, the amygdala goes on reacting to the threats that didn't materialize. Some people become sensitive to a possible danger or threat, and their brain continues to sound the alarm at every hint that that danger might return—even though it never does. The normal pattern is for the mind to cease reacting to things that never happen, through a process known as extinction, but this gradual fading out is tricky to manage because it's hard to tell the difference between avoiding a real threat and avoiding a nonexistent threat. If you fear that Earth will explode unless you eat a doughnut for breakfast, you'll end up eating a lot of doughnuts—hey, it's worked so far, hasn't it?

That particular phobia may be harmless enough (unless you're trying to lose weight), and there are certainly other times when it's useful to have a gut-level fear of threats. The amygdala's negativity bias is so strong that it remains partially protective even after physical injury to the brain, as researchers found when they posed a series of choices to people with partial damage in the amygdala. Sometimes the choice was framed as a loss: Would you rather pay a penalty of 25 cents or make a 50-50 bet that would cost you nothing if you won but a dollar if you lost? (It's smarter to pay the quarter than make a bet that, over the long term, would cost you an average of 50 cents.) How about paying a quarter or making a bet with a 10 percent chance of losing a dollar? (Make the bet.) Even with a damaged amygdala, the people generally made the right choices to minimize their losses. But when the choices were framed in a positive way—Would you rather be *paid* a quarter or make a bet to *win* a dollar?—they became terrible gamblers. They chose long-shot bets and passed up solid opportunities to make money. The brain's ability to deal with good things was so weak that it was disrupted by the partial damage to the amygdala. Only the power of bad was strong enough to survive.

The brain's negativity bias was useful to Felix Baumgartner's hunter-gatherer ancestors, but it wasn't doing him any favors when his heartbeat quickened at the sight of the death row lake or at the smell of the rubber seal in his helmet. It was triggering the fight-or-flight response in his autonomic nervous system, which controls basic functions like heart rate, digestion, sweating, sexual response, and urination. The system operates mostly unconsciously, freeing the conscious mind to concentrate on more novel and grandiose things, like writing a poem or watching a movie or deciding what to do with your life. But the autonomic system will assist even then, such as by adjusting the pupils of your eyes depending on the amount of light. This system manages the body's level of energy, which scientists call arousal (not to be confused with sexual arousal, although there is certainly some overlap). Once Felix became aroused, his adrenal glands pumped stress hormones through his bloodstream. His heart beat faster and he took shorter, quicker breaths. He sweat more. His digestive system slowed down, freeing energy for his muscles—but also giving him a tight feeling in the gut that escalated his panic.

The autonomic nervous system can be aroused by both bad and good things, but researchers have found that bad holds its customary sway. Breathing gets especially short and quick in response to bad things, and the eyes' pupils dilate, another sign of arousal. When people are asked to guess which of two playing cards is higher, the wrong guesses cause more dilation in the pupils than a correct guess. When there's money on the line, people become more aroused by potential losses than gains, and this bias is evident even when they're not consciously fearful. In one experiment, when participants were given a choice between a high-stakes bet and a low-stakes bet, they didn't show any conscious preference: Their prefrontal cortexes presumably realized that the odds were the same either way, so they divided their bets pretty evenly. But their hearts beat faster and their pupils dilated more when they chose the high-stakes bet and the prospect of a bigger loss.

To make the jump from the stratosphere, Felix somehow had to master his gut fears, to control his amygdala, and short-circuit the fight-or-flight response in his autonomic nervous system. He'd thought he could do it himself, but he needed help from someone who understood the mind thing.

The Guru of Good Thinking

Three months after the tearful phone call in the airport, Felix returned to face the suit and the skepticism of his team. His trainers were veterans of U.S. Air Force and NASA programs that immediately screened out anyone with this sort of fear. It was one of the first tests given to applicants trying to become high-altitude pilots or astronauts. Sometimes the applicant would don a pressurized suit and helmet, and the tester would leave the room without explaining what was going on or when he'd be back. The cooling system in the suit would be turned off, so it got warmer and more claustrophobic as the minutes ticked by. The testers, watching through hidden cameras, would leave the candidate there for at least an hour—or until they saw him start freaking out.

Another test was to put the candidate inside a small dark box and leave him curled up there for twenty-four hours. The grand old man among Felix's trainers, Joe Kittinger, had slept peacefully during his box test in the 1950s, but he'd seen plenty of his air force colleagues flunk it.

"Either it bothers you or it doesn't," he said.

Kittinger had never known anyone to overcome the fear, and neither had Art Thompson, the project director. When Felix asked him what kind of training the air force and NASA provided for pilots and astronauts who developed claustrophobia, Thompson replied, "They have a really simple solution. They get rid of you and bring in the next guy." That was the point of the Right Stuff: Either you had it or you didn't.

Thompson and his team discussed their concerns the day that Felix returned to Lancaster, at a meeting to which Felix was not invited. He waited outside while they considered the risks of letting him proceed. At the end of the meeting, the verdict was delivered to Felix: The team had lost confidence in him. They didn't think he was ready to go into the stratosphere and doubted he ever would be. The message was confirmed when it was Felix's turn to meet with them, and he found himself sitting alone. Everyone else took seats on the other side of the room. Felix felt embarrassed and furious. Through twenty-five years of skydiving and BASE jumping, he'd been the guy who always delivered, and now no one believed in him anymore. One show of weakness, and his team had fled to the other side of the river.

He didn't feel much better when he sat down with Michael Gervais, the clinical psychologist who was brought in to work with him. Gervais's specialty was in "high-stakes environments." He had worked with heavyweight boxers, navy fighter pilots, and professional hockey and football players. At their first meeting, Gervais pointed to an empty chair and asked Felix to imagine he had a son—which Felix did not—sitting there. How would he explain the situation to this imaginary son?

Felix felt ridiculous, conducting a conversation with an empty chair, but he started talking. The point of this exercise was to put his problem in perspective, to step back from the immediate crisis and explain the big picture to someone he cared about (even if that someone didn't exist). Gervais wanted to see where this mission fit into Felix's story of his life. How important was it to him, really? Was it worth going through the pain? In working with athletes and fighter pilots, Gervais had found that they persevered through fears and took enormous risks because they couldn't bear the pain of *not* going for it: They'd rather break a bone or die rather than live with the memory of a lost opportunity. By the time Felix finished talking to his imaginary son, Gervais was convinced that the will was there, and he promised Felix that he would show him the way.

Gervais needed him to exploit a set of conscious and unconscious strategies called *minimization,* a term introduced in the 1990s by the psychologist Shelley Taylor of UCLA. By then Taylor had already made a name for herself with her studies of how women with advanced cases of breast cancer formed "positive illusions"—beliefs that were unrealistically optimistic but enormously valuable in helping them cope. Some illusions created a sense of mastery over the event, such as by adopting superstitious beliefs about how they could prevent the cancer from coming back. (One woman, for instance, convinced herself that she could prevent cancer recurrence by eating large amounts of mashed asparagus, even though—or maybe because—she couldn't stand the taste of it.) Another mental strategy was to reframe their situation by comparing themselves with women who had it worse. For example, the ones who had a breast removed typically compared themselves with those who lost both breasts, not with those who'd had a lump removed.

Taylor later theorized that bad events, unlike good events, evoke two distinct and in some ways opposite responses: mobilization and minimization. Mobilization cranks you into high gear—*Wake up and deal with this!*—through the unconscious physiological responses we mentioned earlier as well as through conscious efforts to think more deeply and analyze more carefully. Taylor noted that mobilization can differ by sex. Whereas men often use the classic fight-or-flight response by turning aggressive or withdrawing socially, women often react by showering more affection on their children and other loved ones, a response that Taylor called tend-and-befriend.

Mobilization can be useful in the short term, but there's no point in pumping adrenaline and cortisol and other stress hormones indefinitely. Once the shock wears off, there are automatic minimization responses that release endorphins, serotonin, and oxytocin to produce good feelings. Novice skydivers go through a moment of terror when they jump out of the plane, but they're often euphoric by the time they land. The scarier a horror movie is, the better the audience likes it, because they emerge feeling happy and relaxed.

Fortunately for the film industry, this minimization effect doesn't work in reverse: People who sit through a comedy don't leave the theater saturated with anxiety, hatred, or depression. The body did not need to evolve mechanisms to minimize good events. Going through life in a happy stupor wouldn't be a great evolutionary strategy—you wouldn't see the lion coming—but there's no danger of succumbing to perpetual bliss, because the impact of good events is relatively small and wears off quickly. It's the bad event that must be minimized so that its power doesn't cripple you.

Besides dosing yourself with feel-good hormones, you can minimize bad through conscious and unconscious mental strategies. Freud referred to this technique as a defense mechanism, and the idea has held up much better than most of his other theories. Psychologists have amassed experimental evidence that people do indeed deal with bad feelings by using defense mechanisms like displacement (berating their children when they're really mad at their spouse or their boss) or projection (*You're the hostile one, not me*). We're marvelously adept at convincing ourselves that others are to blame for our own mistakes and failures. This self-deception can be destructive, but as Taylor observed in the women dealing with breast cancer, positive illusions can be quite useful in overcoming the power of bad.

To help Felix minimize his bad feelings, Gervais started by asking him to imagine that he was inside the suit and tied down so there was no way for him to escape. What would happen? Felix figured the panic would get so bad he'd have a heart attack.

"You don't panic to death," Gervais said. The suit didn't have that kind of power over him. The threat was mental, not physical, and the way to overcome it was to identify all the triggers, to chart the buildup of anxiety all the way from the apartment in Los Angeles to the dressing room in Lancaster. Gervais was fascinated by the mind's power to create fear. He thought of it as part of the human ability to tell stories, and imagined it evolving in early humans sitting around a fire on the savanna.

Someone would tell a scary story to the group, and the details would be so vivid—the hiss of a snake, the battle cries of an enemy clan, the wild eyes of someone charging with a spear—that the listeners would shudder. The only thing in front of them was the fire, but the mental images were so real that their bodies responded. That was what Felix had been doing to himself on the drive to Lancaster, telling himself stories and causing his body to respond to a dressing room as if it were a torture chamber.

Gervais offered him a new mental image for that trip: a runaway train. Once Felix started worrying in Los Angeles about the suit, he was boarding a train of negative thoughts. As he drove past the Getty Museum and the death row lake, as he saw the neon light in the dressing room and smelled the helmet's rubber seal and heard the visor click shut, the train was accelerating straight toward disaster. Gervais had Felix draw a sketch of the route, complete with a drawing of a train, and monitored his heart rate as he imagined himself at each point along the way.

If Felix's heart rate and breathing quickened at any point, he had to stop and calm himself before going any further. Gervais asked him to rate his level of anxiety from 1 to 10 as he went along. If it rose to 3 when he passed the death row lake, he couldn't proceed until he brought it back down to 1. If it went back up when he reached the dressing room, he couldn't put on the suit until he was calm again.

The goal was not just to diminish his fear but to eliminate it, using two techniques that Gervais taught him: thinking well and breathing well. Instead of regarding the suit as a prison, Gervais told him, think of how special it is. It's the only custom-built space suit in the world.

The NASA astronauts had to choose from one of three standard sizes, but this one was tailored just for you. It's a beautiful uniform. It's your friend, not your enemy. It will give you pure oxygen to breathe and keep your blood from boiling in a place that other humans can't go. It puts you in an exclusive club. The smell of the rubber seal is the smell of success. The suit turns you from a nobody into a superhero.

Felix was understandably skeptical at the pep talk, but this was a well-established technique from cognitive behavioral therapy (CBT), which has emerged as the gold standard for psychological treatment. CBT is essentially a set of strategies for overcoming the negativity bias. It was developed by psychologists who broke with the traditional Freudian method of treating anxiety and depression. Instead of trying to unearth the childhood traumas supposedly responsible for these conditions, the CBT pioneers focused on the present. They found that depressed people suffer from extreme negativity bias in the way they see themselves, the world, and the future. Depressed people focus relentlessly on their weaknesses and failures, ignoring their strengths and dismissing their successes as flukes. They interpret one setback as a fatal mistake and imagine it leading to the worst possible outcome.

The researchers categorized these forms of negativity bias—"negative filtering," "positive discounting," "catastrophizing"—and developed strategies for rationally overriding the feelings. They found that very simple techniques, like writing down your fears and then forcing yourself to consider an alternative interpretation of the situation, could have a major impact. CBT has repeatedly been found to be as effective as drugs like Prozac or Lexapro in treating depression—and without the side effects. No other treatment has been studied so thoroughly and found to be as safe and effective for such a wide range of problems, including depression, anxiety, and other disorders.

Felix was using a CBT technique called the "coping statement," which had been demonstrated to help people overcome their fears. In laboratory experiments, these statements had even been shown to increase people's ability to tolerate physical pain. Gervais kept repeating the positive statements so that they became mantras that Felix could tell himself whenever anxiety started to build. The more that Felix heard them and repeated them, the more credible they seemed. The suit *was* special. When he tried on the suit and looked in the mirror, he started thinking, *Hey, it's pretty cool.*

To build up his endurance, they started off with just the helmet—no suit, so that Felix could feel the air against his arms and avoid that panicky feeling of being cut off from the world. At the first session, Felix's anxiety was up to 8 at the end of an hour, and they called it off. But the next time he lasted longer, and they moved on to sessions with the whole suit as Felix learned a new way to breathe. When the positive mantras weren't enough, when the rational reassurances from the prefrontal cortex were no match for the primal responses of the amygdala, Felix concentrated on taking deep breaths.

It was a conscious tactic to send unconscious signals through the autonomic nervous system. In keeping with the train metaphor, you can imagine this system having a couple of subsystems that function as a throttle and a brake. The throttle is the sympathetic nervous system, which accelerates the pulse and breathing and the rest of the fight-or-flight response. The brake is the parasympathetic nervous system, which produces a counterresponse sometimes called rest-and-digest or feed-and-breed. It was dubbed the relaxation response by researchers who observed it in people meditating and found that it could be activated through deep breathing.

Felix activated it by pressing his hands and feet tightly together, holding his breath for thirty seconds, then taking slow, deep breaths until the anxiety passed and he was ready to proceed to the next step of his routine. Instead of feeling anxious when it came time to put on the helmet, he was calm, and if his anxiety rose from 1 to 3 when the visor shut, it was still manageable at that level. He'd just repeat a mantra—*This makes me a superhero*—and keep taking deep breaths until it went back to 1. Previously he'd felt panicky and short of air when they sealed the helmet and he heard the labored sounds of his own quick breathing. Now, as he took slow breaths, he relaxed so much that he began savoring the experience of filling his lungs with pure oxygen.

"As stupid and simple as it sounds, breathing helps a lot," he said. "Just getting that good air into your lungs changes everything." Within a few weeks, he was spending hours in the suit. When the time came for the big test, a simulation of the mission, he didn't spend a sleepless night worrying about it. The five hours inside the suit seemed to him to pass quickly, and he emerged to an ovation from the team. Fearless Felix seemed to be back in the building.

Training the Fearless Mind

The NASA and air force veterans were amazed to see the transformation in Felix, but Mike Gervais wasn't surprised at all. There was nothing novel about this treatment. He'd achieved similar results with all kinds of clients: business executives deathly afraid of public speaking, NFL players who choked in big games, employees paralyzed at the thought of asking for a raise, claustrophobes afraid to ride an elevator. In Gervais's view, if you care enough to work through the fear, you can do it—and you may not even have to pay for the therapy. Whatever your demon—spiders, heights, public speaking, or any of the anxiety disorders that affect more than a quarter of the population—you can follow the same strategies that Gervais used with Felix:

1. *Talk about it.* You can start off by conversing with an empty chair, as Felix did, but at some point you want to be heard by a real person, whether it's a therapist or someone else you trust. Talking about a fear helps you overcome it, and talking about the progress you're making helps you make further progress.

   When it comes to sharing information, researchers have found, as usual, that there's a sharp difference between bad and good. People are much more likely to recall good events in their past if they shared the news at the time with someone else. (Later we'll have more to say about this research—and the right way to respond to good news from your spouse or a friend.) But telling someone about a bad event in your life doesn't make you more likely to remember it later. This seems odd, because the mere act of reciting something can strengthen it in your memory, but there are a couple of explanations. One is that you paid so much attention to the bad event when it occurred that the recitation doesn't make it significantly more memorable. The other explanation is that the bad event loses some of its power *because* you talked about it. The listener may help you figure out how to solve the problem, or at the very least reassure you that you're not the only one with this trouble. One of the greatest benefits Mike Gervais provides to a client like Felix is letting him know how many other people have struggled with irrational panic and learned to vanquish it.
2. *Map the runaway train.* If, say, you're a claustrophobe who panics in an elevator, the anxiety probably started increasing before the doors closed. Perhaps it was when you started thinking about going into the building, and then it accelerated further when you went in the front entrance, and then again when you walked down the corridor, and then when you saw the elevator. Plot the waypoints in your mind—or on paper, as Felix did.

3. *Throttle the train.* Once you've identified the waypoints on the route to panic, don't proceed beyond any one of them unless you're calm. If you feel the fight-or-flight reaction starting, use the remaining techniques that Felix learned.
4. *Recite your mantra.* Prepare yourself with coping statements: simple sentences that can be repeated in your mind to overwhelm the irrational negative thoughts. Some therapists advise writing them down on cards that you carry with you. The statements should be short and realistic. If you're afraid of public speaking, don't promise yourself that you'll calmly deliver a flawless address. Instead try something like *People will accept it if I'm nervous and stumble occasionally.* If you're afraid of elevators, remind yourself of the facts. *An elevator is one of the safest places in the world. It's getting me where I want to go. My legs can relax instead of climbing all those stairs.* The lines sound silly at first, but they're true and eventually they work. If you're panicking in an unexpected situation, you can fall back on another line that Gervais taught Felix: *You don't panic to death.* Whatever is happening, you can always repeat to yourself: *This too shall pass.*

The mere act of repeating words is calming, as monks have long known. Felix was so impressed with the techniques he learned from Gervais that he began reading about the Buddhist monks at the Shaolin Monastery in the mountains of China, the famed school for martial arts. The monks who developed Shaolin Kung Fu cleared their minds by chanting mantras, just as European monks used the Gregorian chant for less violent pursuits. Scientists have found that when monks and practitioners of Transcendental Meditation repeat mantras like "Om," the heart rate slows, blood pressure drops, and there are hormonal changes in accordance with the relaxation response.

5. *Breathe.* Press your hands and feet together for thirty seconds, then relax and take slow, deep breaths from your diaphragm. "A deep breath is a signal to the body that we're safe," says Gervais. It's also a way for your mind to concentrate on something other than your fear. Gervais suggests training—and reducing your daily stress—with regular exercises, starting with ten deep breaths every day for ten days, then moving up to twenty breaths a day for twenty days.

There are more elaborate strategies for calming yourself, such as cognitive bias modification, which helps people overcome the brain's automatic focus on negative images and thoughts. This therapy has great mass potential, for both children and adults, because people can do it on their own using a computer or a smartphone app that reinforces positive words or images and positive interpretations of ambiguous statements. For instance, they'll see two faces flash on a screen, one hostile and one neutral. By following instructions to look for a letter on the screen, they train themselves not to focus on hostile faces. Psychologists still aren't sure exactly how much good this does for people dealing with real-world situations, like scanning the faces of strangers at a cocktail party, but it usually does seem to lower anxiety, and some research shows that people stay calmer when dealing with a challenge like giving a speech.

People have overcome their fear of heights thanks to a computerized form of cognitive bias modification called interpretation training. The traditional treatment for acrophobia, like other phobias, is exposure therapy, which requires a therapist on hand to coax the person to go up a few steps, look down, and then gradually go higher and higher. But psychologists at the University of Virginia achieved the same effect simply by putting people in front of a computer screen and assigning them verbal exercises.

The acrophobes would type the missing letters in scenarios like this one: "You are on the roof a five-story apartment building. Grasping the railing, you realize you have never been this high up before. Getting off the roof when you need to will be e_sy." Later they'd be asked whether it would easy or difficult to get off the roof.

Like Felix's mantras, it sounds hokey, but that simple act of reading the scenario and answering "easy" had a significant impact. After just three hours of computer exercises, the people were able to climb stairs and tolerate heights just as well as another group that had undergone conventional exposure therapy. Many of them still felt anxious, but, like Felix, they had a new way to think about the situation and deal with their anxiety.

"People with phobias want to avoid panic attacks, but that's not the right goal to start with," says Bethany Teachman, one of the Virginia researchers. "The first goal is to stop *caring* whether or not you have a panic attack. A panic attack is uncomfortable but not dangerous. It's a false alarm—a fear of fear. Once you make the decision to tolerate it, you get a sense of mastery, and eventually the fear loses its power over you and the panic attacks don't come anymore."

However you confront your fear, whether it's with the help of a therapist, a website, a smartphone app, or the five-step strategy outlined above, the ultimate goal, as always, is to improve that positivity ratio so that there's much more good than bad. When your mind is obsessing on a danger, overwhelm it with positive thoughts. When stress hormones start activating the fight-or-flight reaction, use the relaxation response to produce a surge of antistress hormones that leave you relaxed and reassured. Consciously training yourself to accentuate the positive can help you cope with any kind of threat. By learning to deal with his claustrophobia, a danger that existed only in his head, Felix was able to face the quite real dangers of the stratosphere.

Free Fall

The final struggle with the space suit took place with a global audience watching. Felix donned the suit at four A.M. on October 14, 2012, in a trailer on the airfield at Roswell, New Mexico, where reporters and camera crews from around the world had gathered. He sat there placidly breathing pure oxygen and watching videos until it was time to climb in the capsule. His heart rate remained normal, in the sixties and seventies, and didn't exceed one hundred even when the capsule was jerked off the ground by the largest balloon ever to carry a human aloft.

During the two-and-a-half-hour ascent, the worldwide audience grew to more than fifteen million viewers. Everything seemed to be going smoothly until the balloon approached sixty thousand feet and Felix mentioned something about his helmet.

"Phil, check your monitor," Joe Kittinger said. "Phil, check your monitor." Suddenly there was radio silence.

Kittinger was not really giving instructions about a monitor to someone named Phil. His words were the prearranged code signal to shut off the audio feed to the press corps and the public. They wanted privacy for this crisis. Felix had been fiddling with the dial that regulated the heating filaments in his visor, and he wasn't feeling any heat.

"When I exhale, it's fogging up and it's not really clearing unless I inhale again," he said. "So it's the oxygen that clears the visor, not the heat."

That wasn't an issue while he was sitting in the relatively warm confines of the capsule, but an unheated visor would freeze instantly when he opened the hatch and stood on the capsule's step.

"If I'm standing outside and my visor freezes up, I cannot go back into my capsule and I cannot jump," he told mission control. "This is a serious problem."

He'd be unable to squeeze back into the capsule and reseal it, and there'd be no prospect of his visor clearing as he hurtled through a region of the stratosphere that was minus sixty degrees Fahrenheit. Without being able to see Earth below, he wouldn't be able to orient himself to stop spinning out of control, and he wouldn't know when to pull the parachute. If he pulled too soon, too far above the ground, he would drift so long that he'd run out of oxygen. And if he pulled too late . . .

The engineers at mission control began planning how to abort the mission, but Felix refused to consider it. His breathing and pulse stayed calm as he consulted with the skydiving specialists at mission control and worked out a contingency plan for timing a blind jump. The audio feed was turned back on for the public and the press, with no mention of the new risk. When the balloon reached 127,852 feet, an altitude of twenty-four miles and a record for the highest manned flight, Kittinger and Felix went through a forty-item checklist to disconnect him from the capsule's lifelines and turn on the portable power pack. To everyone's relief, the visor's heating system seemed to be working fine. Felix opened the hatch, treating the earthbound audience to a view of the globe below, its blue edge curving against the black of space.

Felix was wearing one hundred pounds of gear—oxygen, cameras, parachutes, power pack—and the exertion of lifting it out of the capsule sent his pulse up to 160. But this was not a panicky fight-or-flight reaction. He stood on the step, gripping the handrails and trying to "inhale the moment," as he later put it. The reporters back at mission control gasped at the vertiginous view of the planet so far below him—how could anyone jump into that void?—but as far as Felix was concerned, the worst was already over.

He saluted and said his final words from the stratosphere: "I'm coming home now." Then, very carefully, he leaned forward and fell.

Thirty-four seconds and six miles later, he became the first human to break the sound barrier under his own power (with some help from gravity). The dull *boom-boom* was heard on the ground by a rancher and the rescue teams. He kept accelerating, reaching a peak speed of 844 miles per hour, Mach 1.25.

After four minutes and twenty seconds, he finally pulled the parachute, setting a record for a free fall: 119,431 feet. He glided down to the desert plateau, landed gently on his feet, sank to his knees, and punched the air in triumph. Then he stood up, and the rescue team arrived to help him take off his helmet for the last time.

"Yes!" he shouted, raising his arms skyward. "I think I just lost 2,000 pounds of weight off my shoulders. I want to hug the whole world." At that moment, he later explained, he was thinking not about records he'd set but about something more important.

"My biggest accomplishment was not the jump itself," he said. "It was facing my demon, which was the suit. I never thought I would be able to overcome the anxiety. Then Mike Gervais completely changed my mind. It's like a miracle."

One of the first things he did after the jump was to celebrate with Gervais. "I've got a toolbox for the rest of my life," he said, and Gervais agreed. Gervais looked on Felix's quest as an inspiration for anyone dealing with fear.

You'll have bad moments at first, just as Felix suffered through the training. But if you confront the fear properly, you can endure it and master it, just as Felix controlled his claustrophobia wearing the suit in a cramped capsule all the way up to the stratosphere. And then you're rewarded with your own version of the vista that greeted him when he opened the hatch.

"Once you extinguish the fear," Gervais said, "you're looking at a whole world of possibility and freedom."

CHAPTER 4

Use the Force

Constructive Criticism

Bad distorts your judgment, but it can also sharpen your wits. As we'll see in this chapter and the next, the power of bad can spur students, help adults flourish, and promote the noblest of causes. To use its force, you first need to understand the impact of criticism—the pain it causes as well as the benefits it provides—and a good place to start is with Stephen Potter.

Potter was a British humorist with a keen appreciation for the negativity effect long before psychologists came up with the term. He realized how useful it could be—and didn't pretend to harness it for any noble cause. Potter's books were compilations of sly social gambits designed to "make the other man feel that something has gone wrong, however slightly." His 1947 bestseller, *Gamesmanship*, introduced a term that he defined as "the art of winning games without actually cheating." He later broadened the concept to the rest of life and added another word to the language, *one-upmanship,* a strategy enabling the ignorant and unskilled to prevail against their superiors.

Suppose you're at a dinner party being dominated by a cosmopolite expounding on his recent travels. Whatever he's talking about—Chinese politics, the Congo rain forest, Peruvian cuisine—you can throw him off his stride without knowing a thing about the place. Any generalization he makes about any country can be countered with a single sentence uttered with serene authority: "Yes, but not in the south."

Potter viewed criticism as an opportunity for one-upmanship as long as it was delivered deftly, in a pseudofriendly fashion. The essence of reviewmanship, as he called his technique of writing book reviews, is "to show that it is really you yourself who should have written the book, if you had had the time, and since you hadn't, you are glad that someone has, although obviously it might have been done better." Again, no expertise is required. You don't need to know a thing about botany, for example, to review Dr. Preissberger's *Rhododendron Hunting in the Himalaya.* You don't even need to read the book. Randomly pick a name cited in a footnote and lament that the author doesn't give credit in the main text "to that impeccable scholar, P. Kalamesa" (whoever that is). Or find the Latin name of a plant not in the book's index and ruefully note, "Dr. Preissberger leaves the problem of *Rhododendron campanulatum* entirely unanswered."

We don't recommend this approach to criticism—and definitely not to anyone reviewing *this* book. But we do admire Potter's insights. He understood the effects of mixing praise and blame better than most parents and executives today. It took researchers a while to appreciate his theory of literary one-upmanship, but eventually it was tested in some clever experiments, like the one in which people were shown a positive book review:

> In 128 inspired pages, Alvin Harter, with his first work of fiction, shows himself to be an extremely capable young American

> author. *A Longer Dawn* is a novella—a prose poem, if you will—of tremendous impact. It deals with elemental things: life, love and death, and does so with such great intensity that it achieves new heights of superior writing on every page.

How intelligent does the critic seem to you? How likable? Now consider this review:

> In 128 uninspired pages, Alvin Harter, with his first work of fiction, shows himself to be an extremely incapable young American author. *A Longer Dawn* is a novella—a prose poem, if you will—of negligible impact. It deals with elemental things: life, love and death, but does so with such little intensity that it achieves new depths of inferior writing on every page.

The syntax and analysis in each review are identical, but replacing each positive word with its opposite does wonders for the critic's reputation. In the experiment, people who saw the negative version of the review rated the critic as significantly more intelligent than did the people who read the positive version. The naysayer got lower ratings for kindness, fairness, and likability, but he was rated higher for literary expertise. Teresa Amabile, the psychologist who did the experiment, also tested this effect by using two reviews that had been published in the *New York Times Book Review*—one a rave, the other a pan. Both were written by the same critic, but Amabile disguised that by changing the bylines when she asked people to rate the author of each review. Sure enough, the author of the scathing review seemed smarter than the author of the rave, even though it was actually the same person. Amabile concluded that when you're trying to make an impression, you often face the choice of seeming "plodding but kind" or "brilliant but cruel."

What you choose is strongly influenced by the situation, as Amabile demonstrated in another study. Each person in the experiment was put in front of an audience and asked to evaluate someone else's work. If this critic believed herself to be higher in status than the audience and was feeling intellectually secure, then she was often charitable in her evaluation. But if she was feeling intellectually insecure and lower in status than the audience, then she was much more likely to go negative. She would try to raise her status by employing the brilliant-but-cruel strategy. Being caustic wouldn't win her friends, but it could earn her respect even if the criticism was unfair. It's the sort of upmanship practiced by Elizabeth Bennet, the heroine of Jane Austen's *Pride and Prejudice,* when she meets a social superior, the wealthy aristocrat Mr. Darcy. She immediately starts mocking him behind his back. Eventually, after realizing her injustice to him, she confesses to her sister why she did it.

"I meant to be uncommonly clever in taking so decided a dislike to him, without any reason," she explains. "It is such a spur to one's genius, such an opening for wit, to have a dislike of that kind. One may be continually abusive without saying anything just; but one cannot be always laughing at a man without now and then stumbling on something witty."

To seem brilliant yet not cruel, you can try tempering the criticism with kindness, or at least the faux kindness taught by Potter. One of his favorite ploys was to preface a negative comment with "I'm afraid," as in "I'm afraid that Dr. Preissberger's prose makes for some rough sledding." Potter called the strategy *I'm Afraidmanship* and declared it "admirable for showing that you are a nice man." Once again, Potter's wisdom has been confirmed by experimenters, who have their own name for this stratagem: the *dispreferred marker.* That's the linguistic term for a phrase like "*I'm afraid*" because it serves as a signal (a marker) that the sentence includes something negative (dispreferred). We often use these markers at the start of a sentence, as with *To be honest* or *With all due respect* or *Sad to say.* But they can be dropped in anywhere to cushion the blow: "By the 17th rhododendron-hunting trek into the mountains, most readers will be exhausted, but Dr. Preissberger, *God bless him,* perseveres for another 200 pages."

To test these phrases, a team of researchers led by the consumer psychologist Ryan Hamilton showed people listings for products from Amazon.com along with a customer's review that was generally favorable ("This watch took my breath away") but also included some criticism at the end ("the band can sometimes pinch and rub"). Other people saw a version in which a few extra words—*I don't want to be mean, but*—were inserted before the criticism. Those few words made a difference, and so did other dispreferred markers like *I've got to be honest* and *Don't get me wrong.* The readers became more eager to buy the watch than if they'd seen the unedited review. They knew the watch wasn't perfect, but they overlooked its flaws because they found the reviewer more likable and also more credible. When it came time to publish the results in the *Journal of Consumer Research,* the researchers used the stratagem themselves. They gave the article a proper scholarly subtitle, "The Use of Dispreferred Markers in Word-of-Mouth Communication," but they prefaced it with "We'll Be Honest, This Won't Be the Best Article You'll Ever Read."

While being a critic makes you look smarter, what if that's not your chief objective? What if, unlike Potter, you actually want to offer constructive criticism? That's a trickier proposition. Delivering criticism or other negative messages can produce lots of positive results, but only if it's done correctly, and most people don't know how.

Delivering Bad News: The Wrong Way

For too long, business executives have been force-feeding their subordinates the "criticism sandwich." The idea, also called the "feedback sandwich," was popularized in the 1980s by Mary Kay Ash, the founder of Mary Kay Cosmetics, who advised managers to sandwich any critical remarks between layers of praise. It sounds logical enough, and it makes the annual evaluation less painful for the manager. Giving criticism face-to-face is difficult for most people, so it's more pleasant to start with the good stuff. The manager goes on at length about the employee's strengths and achievements before getting to the meat of the criticism. Then she switches back to conclude with a few nice words and end on a happy note—or so it seems to the manager.

But that's usually not how it feels to the employee. By this time all the opening praise has been forgotten. The employee can't get the bad stuff out of his mind. He's choking on the middle of the sandwich. A conversation that was supposed to inspire better work has left him demoralized.

This problem, of course, is not limited to annual evaluations. It's one of the oldest and most awkward social conundrums: How do you deliver bad news? When Douglas Maynard, a sociologist at the University of Wisconsin, began systematically analyzing this question, he noted the answers that have evolved in folklore over the centuries. One view is found in a joke about the man who takes care of his sister's beloved cat while she's traveling. A week into her trip, he calls her and promptly delivers the bad news: "Your cat is dead." The sister is devastated and rebukes him for his abruptness. He should have broken it to her gently, she says, suggesting he could have first called to say that the cat was on the roof of the house with no way to get down safely.

"Then we would hang up," the sister explains, "and you would call a little later and say that the cat had fallen off the roof and was injured. And then you could call again to say the cat had died. That way I could have gotten used to the idea." Her brother apologizes, and she continues on her trip. A week later she gets another call from him.

"Hello," he says. "Um, Mom's up on the roof, and we can't get her down."

This joke plays off a common assumption that bad news should be delivered gradually because people aren't ready to hear it. But do they really want to put it off? Not according to researchers who have tested that old line "I have some good news and some bad news" by asking people which they want to hear first. More than three-quarters of the people want to start with the bad news. It's only when asked how they'd prefer to deliver mixed news that people say they'd be more comfortable starting with the good news. A manager thinks he's being kind by starting off with heaps of praise, but he's doing it more for his own benefit. Most employees would rather first get the bad stuff out of the way.

In fact, the opening praise can make the subsequent criticism even more painful, as Baumeister found in a joint study with the clinical psychologist Kenneth Cairns. In the study, college students took a personality test and then were given feedback that was supposedly generated by a computer analysis of their personality. (The feedback actually had nothing to with the students' responses, but that white lie induced them to pay more attention to it.) The computer flashed a list of adjectives that supposedly fit the student's personality, and afterward the student was asked to write down as many as he or she could remember.

Some of the students saw a list containing mainly positive adjectives like *confident* and *honest* along with just a few negative words like *spiteful* or *greedy;* others saw a list of mainly negative words. No matter which list they saw, the words of praise didn't make much impact, because the students recalled fewer than half of the positive adjectives. The criticism had more impact, but not in every situation. When the students saw a list of mainly negative adjectives, they forgot most of those words, too.

It was only when they got the equivalent of a criticism sandwich—a list of mainly positive adjectives that included just a few negative ones—that the bad stuff became truly memorable. Most of those isolated zingers stayed in their memories. The explanation for this became clear when the researchers analyzed the different ways that criticism was processed. Based on the prior psychological tests, they knew some of the students were already experts at denial. They met the definition of a *repressor,* meaning someone who tends to ward off or deny anything bad. (Clinical psychologists consider repressors especially difficult to work with, because they won't admit they need any help, and managers may find them to be difficult employees, too.) In the experiment, the repressors were quite adept at ignoring a litany of criticism. When they saw the list of mainly negative adjectives, they zipped through them, quickly tapping the keyboard to move from one word to another, and afterward they were less likely than the other students to remember the barbs. No surprise there—that's what repressors do.

But a remarkable reversal occurred when the repressors were shown the mainly positive list. They relaxed their defenses, spending more time to enjoy each bit of praise—and then were staggered when an adjective like *hostile* or *deceitful* appeared. Afterward, they were even *more* likely than the other students to recall each of those flaws. As expert as the repressors were at dodging bad news, they turned out to be especially vulnerable to the criticism sandwich. And even though the rest of the students—the nonrepressors—weren't hit quite as hard, they, too, remembered those isolated bits of criticism better than any other words in the experiment. The experiment's results jibed with Thomas Jefferson's observation: "I find the pain of a little censure, even when it is unfounded, is more acute than the pleasure of much praise."

For praise or other good news to make a lasting impact, the brain must transfer it from the short-term working memory into long-term memory. This process gets disrupted when the good news is followed by something negative. The brain uses so much energy to focus on the new threat that the previous pleasantness gets lost because of an effect called retroactive interference. Just how it's lost is a subject of debate among researchers; some think it's actually erased from memory, while others believe it's still there but difficult to retrieve because of competition from the new negative information. Either way, it explains why people often can't recall what they were doing just before something bad happened. It's why so many employees walk out of meetings obsessing about the one or two bits of criticism they got instead of all the praise that preceded it. The criticism sandwich may be logical, but the brain doesn't logically process threatening information. The power of bad can short-circuit the ability to remember good. When you have something painful to say, you need to take the negativity effect into account—and then put it to work for you.

Delivering Bad News: The Right Way

To deliver criticism or bad news, first know your audience. That may seem like an obvious first step, but it's routinely skipped, and not just by the managers who have been feeding everyone the same criticism sandwich. Doctors ought to be the world's foremost experts at delivering bad news—they do it all day long—yet too many of them haven't mastered this basic step. There's a vast medical literature on how to break bad news to patients, but much of it presumes that doctors should set the agenda. They're supposedly so knowledgeable and powerful that they can plan in advance exactly how to structure the conversation and control their patients' reactions.

But when the sociologist Douglas Maynard studied these conversations, he found that the doctors with the best bedside manners didn't set the agenda or follow any script. Instead, they took their cues from the patient. These doctors made sure to deliver the news in person, not by phone or email, so they could observe the patient's reactions and adjust accordingly. They'd frequently start off by asking the patient how he was feeling. Then they would seek the patient's perspective by asking a question like "What have you learned so far?" or "What do you think is going on?" If the doctor had a written report with test results, she might let the patient read it and then ask, "What do you think this means?"

Asking questions allows the patient to be more than a passive audience. The first impulse when gobsmacked with bad news is self-protection: the fight-or-flight response. Some patients try to shut out the news; others want to shoot the messenger, or at least argue with her. But if the patient is instead asked his perspective and is the first one to say that something's wrong, then he's readier to face it and continue the conversation.

Once he acknowledges that there's a problem, the doctor can confirm his observation and explain why he's right. Maynard calls this the "perspective display sequence," a three-step process in which the doctor first seeks the patient's perspective, then confirms it, then delivers the details of the bad news. Instead of being the hated bearer of ill tidings, the doctor becomes someone who agrees with the patient and wants to work together to deal with the problem. Of course, it doesn't always work this way. When asked for their perspective, typically a third of patients will deflect the question and say something like "You're the doctor. You tell me." But even then, the patient feels like more of a partner, and the doctor has gained a better sense of how the patient wants to deal with the news.

As the doctor proceeds with her explanation, she can keep watching the patient and asking questions, taking care not to go too fast. After delivering bad news, there's a temptation to rush ahead to something less painful, like offering encouragement or focusing on the logistics of handling the problem. But the best way to follow bad news is to shut up. A pause gives the patient a chance to absorb the blow, and it allows the doctor to gauge the patient's reaction. If the patient is speechless, the doctor can gently coax a reaction—"I know this must be hard"—and ask questions like "What worries you the most about this?" Instead of proceeding with more explanations and plans, the doctor needs to figure out what the patient needs next.

The same guiding principle applies to delivering any kind of bad news or criticism. Here are some strategies:

Consider your objective. Do you simply want to help someone cope emotionally with unpleasant facts, or are you trying to spur them to change? In either case, the person will feel better if there's some good news along with the bad, but the sequence depends on your goal. In a study of sequencing strategies, the psychologists Angela Legg and Kate Sweeny administered a personality questionnaire to people who were then given feedback (bogus, as usual) about their traits. When the people heard first about their bad traits and then their good ones, they ended the experiment in a better mood, but they were also less inclined to do work to correct their bad qualities. The ones who heard the bad traits last were more worried but also more eager for self-improvement. It's not easy to motivate without demoralizing, but you can compromise by concentrating on the good feedback toward the end while also finishing up with a clear reminder of what's wrong and how to fix it.

Ask questions. They're useful whether you're critiquing a romantic partner, a friend, a student, or a colleague at work. If you're doing a formal evaluation of an employee, don't automatically serve the criticism sandwich or any other dish. A better culinary metaphor would be to offer a menu.

You can start off with a quick welcome, letting the employee know that he has done some good things and you're looking forward to even better work the next year. A little positive reinforcement makes people more receptive to subsequent negative feedback, researchers have found, and you don't want him sitting there wondering if this meeting is going to end with him being fired. But then see how he wants to proceed. Offer him a menu of the topics to be covered: his strong points as well as the not-so-strong points and the ways to improve. As we mentioned earlier, most people prefer to get the bad stuff out of the way first, but if this employee isn't one of them, proceed with some praise. Just don't go on too long, and don't assume he'll remember it. You'll have to repeat it later, because it will register only after you've gotten through the bad stuff.

You can ease into the criticism with the sort of question used by doctors delivering bad news: "How do you think things are going?" If there's a written record to consider—productivity or sales figures, a list of projects completed—you could ask the employee to look at it and offer his judgment. Some employees (like those repressors studied by Baumeister) will refuse to identify any weakness, forcing you to reveal to them that they're not perfect. But most will probably recognize some area for improvement, and there's your opening. You can confirm the employee's judgment and then gently expand the discussion to offer your analysis of his problems. As you offer criticism, be sure to give the employee time to absorb it, and monitor his reactions with more questions like "Is that fair?" or "Does that make sense?"

Once you've gotten the criticism across, use the power of bad to your advantage. As soon as the brain receives criticism, it kicks into high gear, alert for information to deal with the threat. Now you can start accentuating the positive, because the good stuff will be encoded in long-term memory—along, of course, with the bad stuff. As usual, it will take a few good things to make up for every bad one, so don't stint on the compliments. One way to improve the positivity ratio is to focus on future achievements rather than past mistakes. You can point out that the employee missed some deadlines in the past year, but then spend lots more time discussing how to structure the job so that he succeeds in the coming year. If the employee's personality got him into trouble on team projects in the past year, make plans for more solo projects this year, and praise him as a self-starter who excels when he has the freedom to operate on his own.

In doling out praise, don't worry that it will seem overblown or insincere. Most of us like to think we see through false flattery, but that's not what the sociologist Clifford Nass found when he offered it in his lab at Stanford. He programmed a computer to brownnose people playing a game of Twenty Questions. The computer would ostensibly think of an animal, and each player would try to guess it after asking a series of yes-or-no questions. The computer never told the players whether they'd guessed the right animal or not, but it did praise them for asking questions that were "ingenious," "clever," and "highly insightful." Some of the players believed the praise was deserved, because they'd been told that the computer really was evaluating how well they played the game. These players were naturally quite satisfied with their performance. They were also satisfied with the computer, giving it high marks for both likability and accuracy.

Other players, though, were warned at the start of the experiment not to believe the computer. The evaluation software wasn't yet finished, they were told, so for now the computer was merely offering stock comments that had nothing to do with how well the players performed. The players, who happened to be computer-science students, understood the situation clearly. When asked afterward if they'd paid any attention to the computer's comments, they all said no. One of them indignantly wrote, "Only an idiot would be influenced by comments that had nothing to do with their real performance."

Yet the flattery *still* worked. It bolstered the players' opinion of their own performance as well as the computer's. They liked the computer just as much as the players who'd been told the praise was genuine. They even gave the computer equally high marks for accuracy. Although they were consciously aware that the praise was bogus, they unconsciously absorbed the flattery and liked the flatterer. The research convinced Nass that flattery just about always works as long as there's enough of it.

Be creative with your praise. To counterbalance a few negative comments in a meeting, Nass advised, you should come prepared with a long list of positive comments, ideally ones that are memorable because they're surprising or clever. Instead of just praising a salesman's skill, call him "The Closer." Such flattery might seem gratuitous—or unctuous—when you're dealing with someone who's obviously talented and has every reason to be confident. But don't assume that success offers any protection against criticism. When Ronald Reagan was president, he visited New York one day and was cheered by tens of thousands of people lining the streets as he rode through midtown with Mayor Ed Koch. As they were crossing Forty-Second Street, Reagan looked out the car window and exclaimed, "Look at that guy—he gave me the finger!" Koch couldn't understand the reaction.

"Mr. President," he said, "don't be so upset. Thousands of people are cheering you and one guy is giving you the finger. So what?"

"That's what Nancy always says," Reagan replied. "She says I only see the guy with the finger."

No one is immune to criticism. Lee Daniels has directed an Oscar-winning movie (*The Butler*) and created a hit TV show (the critically acclaimed *Empire*), but he still can't bear reading anything negative about his work. He'll read a rave review and obsess about the single sentence of criticism. "It's like taking a knife and stabbing you in the heart over and over," he says. "So I've learned to protect myself by not reading reviews."

No matter what line of work you're in, we recommend using Daniels's strategy: *Let someone else read your reviews.* (And make sure it's someone who won't let that killer sentence get through.) The rise of social media means that just about every business or profession is now fair game for critics. You can't ignore them altogether, and you shouldn't—there may well be some useful guidance from customers on Yelp or students on RateMyTeachers. But if you scroll through all the vitriol yourself, you'll get so obsessed by the nasty insults and cheap shots that you won't remember anything else.

Better to let someone else read through the reviews, pick out the fair criticism, and then give you a synopsis along with some excerpts—a carefully edited mix of positive and negative comments. Your personal censor should be kind enough to follow up any criticism with lots of praise, but also tough enough to recognize how useful the bad stuff can be. "Criticism," Winston Churchill observed, "is like pain in the human body. It is not pleasant, but where would the body be without it?"

Like physical pain, criticism is essential because it focuses attention on something that's gone wrong and is liable to get worse. When delivered properly, with the right mixture of praise, a few negative words are often powerful enough to correct the problem. Sometimes, though, it takes a more potent form of bad.

CHAPTER 5

Heaven or Hell

Prizes vs. Penalties

Whether you're trying to improve yourself or someone else, criticism does only so much. If it's not producing results, or if you're looking to avoid future problems, you need to use direct incentives, positive or negative. That means considering one of the oldest questions in social science, and also one of the most dubious clichés: The carrot or the stick?

Lexicologists have traced the expression to the mid-nineteenth century, when cartoons in Europe and America depicted a jockey winning a race by dangling a carrot in front of his mount, and a widely reprinted story used the tale of a donkey that responded only to a reward, not a blow from a stick, as a lesson in child rearing. This purported bit of folk wisdom was summarized in 1851 in an American magazine: "All nature says, 'Lead! don't drive!' from the experiment of the carrot-persuaded donkey."

But did anyone ever conduct this experiment? We've never noticed a carrot dangling in front of a horse in the winner's circle at the Kentucky Derby. Jockeys prefer a stick approach, the whip, which is the traditional motivational tool used with donkeys and mules, too. If those nineteenth-century moralists had bothered watching the muleteers of their day, they wouldn't have spotted any enticing vegetables. The men in charge of artillery wagons during the Civil War didn't rely on kindness to coax their mule teams up muddy hills. They were renowned for their skill with the whip as well as for their creative profanity, which was tolerated by their superior officers because the muleteers insisted that cursing at the animals was an essential part of the job.

So we draw a different lesson from the old carrot-or-stick stories. What they really illustrate is our desire to *believe* that rewards work better than punishment. Dangling a carrot is more pleasant than wielding a stick, so we tell ourselves that it's more effective even when there's contradictory evidence. This belief persists in today's parents, teachers, and managers even as researchers keep confirming the power of bad in their experiments with incentives. It also persists among another group of professionals, the clergy, who have been conducting a much longer-running experiment in the motivation of virtuous behavior.

For more than three centuries, Christian preachers in America have been testing solutions to a perennial problem: how to fill the pews on Sunday morning. This wasn't an issue among the zealous Puritans who founded the Massachusetts Bay Colony, but their devotion did not endure. By the early eighteenth century their descendants were an impious people, and so were the rest of the colonists. Only one in five Americans belonged to a church. Americans downed six drinks per day on average, typically starting at breakfast, and there were more taverns per capita in Boston and Philadelphia than in Amsterdam. Judging from the number of babies born less than nine months after the wedding day, more young couples were engaging in premarital sex than going to church. Clergymen regularly lamented the descent of the colonies into godlessness.

Yet when the British preacher George Whitefield traversed the colonies in 1740, he caused a sensation along the Eastern Seaboard. From Georgia to Maine, thousands of farmers would abandon their fields to go hear him preach in a meadow or in the nearest town's square. In Boston he drew a crowd of thirty thousand—more than the entire population of the city. Whitefield's listeners would shout in anguish, break down in tears, proclaim themselves born again, and go home vowing to sin no more.

"It was wonderful to see the Change soon made in the Manners of our Inhabitants," Benjamin Franklin reported in Philadelphia. "From being thoughtless or indifferent about Religion, it seem'd as if all the World were growing Religious; so that one could not walk thro' the Town in an Evening without Hearing Psalms sung in different Families of every Street."

Why did America suddenly get religion? Historians have struggled to explain the First Great Awakening, as they call the revival led by Whitefield and American preachers like Jonathan Edwards. Theories have variously linked it to cultural shifts, social turmoil, and economic fears. But the simplest and most convincing explanation comes from two sociologists, Roger Finke and Rodney Stark, who study the rise and fall of churches—and have independently confirmed the negativity effect.

Hell on Earth

A religious denomination begins as a sect, a small group of passionate believers who set themselves apart from society, like the Puritans who fled England because they refused to conform to the establishment

Anglican Church. Once the Puritans settled in Massachusetts, they were no longer outsiders. They *were* the establishment. The sect grew into the Congregational Church, which dominated New England in the same way that other colonies were dominated by the Episcopal Church (the American branch of Anglicanism). Those mainline churches received government subsidies and could survive without attracting passionate new members.

Their clergymen were well-educated gentlemen, not charismatic rabble-rousers. They preached elegant, cerebral sermons based on the theology they had studied at Harvard and Yale, where rationalism was prized and emotionalism disdained. They had been taught to see God as distant and abstract, a vaguely benevolent deity nothing like the wrathful figure in the Old Testament who condemned sinners to perdition. The sophisticated modern clergyman did not use the pulpit to thunder about eternal damnation. He didn't necessarily even believe in hell.

The revivalists did—most emphatically. George Whitefield told his American audiences not to be lulled by modern theologians who denied "The Eternity of Hell-Torments," as he titled a sermon in Georgia. "Woe unto such blind leaders of the blind," he said, warning that their denial of hell was the surest way to "promote infidelity and profaneness." He urged sinners to imagine themselves forever tormented by "insulting devils" and "everlasting burnings" and the "never-dying worm of a self-condemning conscience." That scene was elaborated by Jonathan Edwards in his famous sermon of 1741, "Sinners in the Hands of an Angry God." Forsaking the theology he'd been taught at Yale, he compared his listeners in Connecticut to a "loathsome insect" dangling over the pit of hell.

"O sinner! Consider the fearful danger you are in. 'Tis a great furnace of wrath, a wide and bottomless pit," Edwards warned. "You hang by a slender thread, with the flames of divine wrath flashing about it, and ready every moment to singe it, and burn it asunder."

Those sermons appalled the theological establishment. An association of Congregationalist ministers denounced Whitefield for using "his utmost Craft and Cunning, to strike the Passions and engage the Affections of the People." Harvard's faculty charged him with the crime of "Enthusiasm." Ezra Stiles, a Congregationalist minister who went on to become Yale's president, complained that the revivalists' strategy was to drive people "seriously, soberly, and solemnly out of their wits." The mainline churches used their political power in some places to prevent the revivalists from preaching, but it was a losing struggle, especially after the American Revolution produced a nation that did not recognize any official religion. Once the mainline churches lost their privileged status (and subsidies), their preachers had to compete, and the competition was hell—literally.

As a motivational strategy, fire and brimstone prevailed during the First Great Awakening and long afterward. Since the eighteenth century, the rate of church membership has tripled in the United States, which is remarkable by contrast with the centuries-long secular trends in Europe. Why do two-thirds of Americans today belong to a church while so many pews in Europe are empty? In their incisive sociological analysis, *The Churching of America,* Finke and Stark conclude that it's not because Americans are an inherently spiritual people or suffer from peculiar cultural anxieties. The difference is that while European governments have continued to officially recognize and subsidize the establishment churches, the United States hasn't given any church a monopoly.

Once competition began in the eighteenth century, the greatest surge in devotion occurred not in mainline churches but in new Methodist churches continuing the hellfire tradition of Whitefield and Edwards. The Methodist preachers, far from being products of divinity school, were often local residents, unpaid amateurs supervised by visiting circuit riders who themselves lacked seminary training. From a tiny sect in the 1700s, the Methodists grew by 1850 into America's largest religious denomination—and then they ran into the familiar problem of mainline churches. As the Methodist Church prospered, it established seminaries whose graduates came to preach a gentler message known as the "New School." Traditionalists complained that the "characteristic idea of this system is benevolence."

Once again, the benevolent message could not compete with hell. By the end of the nineteenth century Methodists were no longer the largest religious denomination in America. The newly triumphant upstarts were hell-fearing Catholics and Baptists, whose churches grew quickly into the twentieth century. Eventually many of their clergy modernized their message, and they, too, lost ground to revivalist preachers, this time to the evangelical and Pentecostal sects that grew so rapidly in the 1980s and beyond. As always, the establishment complained about the upstarts' crude theology, but in the 1980s one mainline clergyman, Bishop Richard Wilke, urged his fellow Methodists to learn from the competition.

"The churches that are drawing people to them believe in sin, hell, and death," Bishop Wilke explained. "Jesus, who knew what he was talking about, explained them, experienced them, and conquered them. If there is no sin, we do not need a Savior. If we do not need a Savior, we do not need preachers." Without evil and the threat of hell, preachers would be out of luck, out of relevance, and out of a job.

The history of Christianity in America isn't a controlled experiment, but the data set is impressive: hundreds of millions of people exposed to competing incentives. They must have enjoyed hearing sermons about a kinder, gentler deity, but what actually packed them into the pews were the threats from an angry God. And this effect has been demonstrated in more controlled conditions by the psychologist Azim Shariff and his colleagues.

In one experiment, Shariff gave students a math test and asked them not to cheat on it by exploiting a computer glitch. Naturally, he'd rigged it so that he could tell if they did. When he looked for factors that distinguished cheaters from noncheaters, he found that it didn't matter much whether the student was male or female, or how the student scored on a personality test. It didn't matter if the student was religious or not—the devout students were just as likely as the others to cheat. What did matter was how the students conceived of God. If they described God with words like *vengeful, fearsome, punishing,* and *angry,* then they were significantly less likely than average to cheat on the test. But if they used words like *forgiving, comforting,* and *loving* to describe God, then they were more likely to cheat. Jonathan Edwards would have approved of the title of the resulting paper: "Mean Gods Make Good People."

Looking far beyond the lab, Shariff compared crime rates in dozens of countries. His regression analysis showed that a country's rate of homicide (the most reliable barometer of crime) is much less strongly related to the level of poverty or income inequality than it is to religious beliefs, as reported in surveys of nearly 150,000 people in sixty-seven countries. These surveys included separate questions about heaven and hell. In some countries, many more people believe in heaven than in hell—and in those countries the homicide rate tends to be higher than average. The promise of celestial rewards apparently doesn't do much to deter murderers in those places. In other countries, though, there does seem to be a supernatural deterrent, because the higher the belief in hell, the lower the homicide rate. The power of bad extends from this world into the next, at least when it comes to restraining people's violent impulses: Hell is stronger than heaven.

Reward and Punishment

Three centuries after George Whitefield, experimental psychologists had their own great awakening. This was in the 1950s, when the study of animal learning was the main business in just about all American psychology departments. B. F. Skinner's methods and other forms of similar learning by association were dominant. One of the universally accepted principles was that reward and punishment had to be immediate. To teach an animal not to do something, it had to be punished right after its transgression—within half a second, according to the best estimates of the day.

But then a young psychologist named John Garcia reported something strange. Working in a U.S. Navy lab in San Francisco, he was using rats to study the effects of radiation, a topic of keen interest in those early days of atomic weapons. Rats were put in chambers and exposed to various levels of radiation. They had plastic water bottles in there, but Garcia observed that they soon became reluctant to drink from those bottles. He wondered whether they might start to associate the taste of water from those plastic bottles with the sick feeling they got from the radiation. (He himself was well acquainted with nausea. In World War II, he had been a pilot in the Army Air Corps, but he'd had to give up flying after being plagued with motion and altitude sickness.) Garcia started testing rats' responses to various tastes, sights, and sounds. He let some rats drink sugary water, which they loved, before going into the radiation chamber. Later, after exposure to radiation, they no longer wanted any more of the sweet beverage.

Garcia's findings caused a sensation. Learning by association was supposed to take multiple rounds of reinforcement (the famous "learning curve"), but Garcia's rats would avoid the sweet-tasting water after just one bad experience. Moreover, learning was supposed to ensue only if the reinforcement occurred within half a second of the experience, but Garcia's rats learned to hate the sweet water even if the negative reinforcement—the nausea experienced in the radiation chamber—didn't occur until several hours after drinking the water. Many experts thought Garcia's results must be wrong, or possibly fake, but further studies confirmed what came to be known as the "Garcia effect." Rats really could learn even when the reinforcement came hours later—but only if the reinforcement was negative. They could learn to dislike a food that made them sick a few hours later, but they wouldn't come to like a food that later produced good effects.

Did human brains and taste buds show a similar bias? Researchers in Belgium let high-school students sample a series of liquids containing neutral flavors that were mixed either with sugar (which the students liked) or with a very bad-tasting, bitter beverage. A week later, they tasted and rated more liquids with the various neutral flavors, but without the sugar or the distasteful mix. This time, they disliked the flavors that had previously been given with the bitter mix, but they didn't show any affection for the flavors that had been mixed with sugar. Like the rats, they had learned from their bad experiences, not their good ones.

This effect explains why people will forever swear off one type of drink, like rum or margaritas, after one nauseating experience in their youth. Baumeister observed a similar phenomenon in his young daughter. As a toddler she was such an adventurous eater that she even liked raw fish. One time, when offered baby food while the adults were having sushi, she protested, "I not hungry for baby food. I hungry for sushi." At age four, she served as a flower girl at the wedding for one of the graduate students in her parents' lab. Sushi was served. The various graduate students, friends of the bride, were amazed that a little girl liked sushi, and they kept offering her helpings to see how she happily ate them. But then it was too much, and she felt a bit sick. After that day, she was never again hungry for sushi.

One of the most elegant studies of rewards and penalties involved young children and marbles. The children in the experiment were being taught to control their behavior, and the marbles were their incentive. Some of the children started out with an empty jar, and then received a marble each time they responded correctly. The other children started off with a jar full of marbles and lost a marble each time they made a mistake. The rewards and punishments were exactly balanced, so if two children learned at the same rate, they would leave with exactly the same number of marbles, regardless of whether they started off with an empty or a full jar. But there was a marked difference in how they learned. The students who were penalized for mistakes learned much more quickly, so they ended up with more marbles than the ones who learned by being rewarded.

Parents and teachers worry that penalties will sadden children, but the psychologist Joseph Forgas has found surprising advantages to being in a bad mood. Your memory improves, as demonstrated in experiments asking people to recall items they saw in a store or to describe events. The people in bad moods were more accurate, apparently because they were paying more attention, and they were also more succinct, providing the relevant information without throwing in redundancies or irrelevancies. When asked to judge the quality of someone's work, they were less biased by superficial impressions of what the person looked like. In experiments testing the ability to spot liars, the unhappy people performed better than happy people. A bad mood made them less gullible.

Failing feels bad but provides more information than success. Trying to understand why you got a good grade on a test isn't all that informative because you had to do well on every part of it. But analyzing a bad grade forces you to zero in on what you got wrong. In European military history, the strongest armies have repeatedly turned out to be the ones who lost the previous war, because the defeat inspired them to reorganize and make strategic innovations while the victors remained complacent. A reward focuses you inward; a penalty forces you to look at the world more carefully and make changes.

The adaptive benefits of bad have been demonstrated in lab experiments that trained people to anticipate irritating blasts of noise. If they could do something to prevent the noise, they paid more attention than if the noise was unavoidable. These benefits have also been demonstrated in studies of people's incentives to make changes. Most people would like to lose weight, but what primarily motivates them to take action is the fear of looking fat, not the hope of looking slim. Most people want to help others, but they're more likely to do it in response to a negative rather than a positive stimulus. In one field experiment, researchers working with the Red Cross sent out two slightly different appeals for blood donations. The positive one, which asked the recipient to help save someone's life, motivated some people to show up for the local blood drive. But 60 percent more showed up after receiving the negative appeal to prevent someone from dying. Dozens of other studies have shown that people, especially women, are much likelier to respond to public-health messages—like appeals to be tested for cancer or heart disease—if the message is based on fear. As Dr. Johnson observed, the prospect of death concentrates the mind wonderfully.

This evidence has not diminished the popular desire to believe that carrots work better than sticks. Instead of recognizing the value of penalties, parents and educators have been moving in the opposite direction, away from the harsh punishments that used to be the norm, and the result has been disastrous for many students, particularly boys from disadvantaged homes. The education establishment, like the mainline religious establishment, has gone on preaching benevolence despite its ineffectiveness. Teachers are trained at education schools to focus on praising children rather than highlighting their mistakes, and schools are reluctant to penalize either teachers or students for failure.

This philosophy is due partly to the self-esteem movement of the 1970s, one of the sorrier mistakes in the history of psychology. Researchers noticed that high self-esteem correlated with personal success in many endeavors, so they concluded that promoting self-esteem would help students thrive. Those researchers, unfortunately, had the causation backward: Being successful will indeed raise your self-esteem, but having a high opinion of yourself will not make you more successful. The self-esteem theory was soon enough discredited among research psychologists, but it has persisted among many educators and supposed experts in child development. The establishment consensus, as spelled out in a widely used guide for teachers titled *Best Practice,* urges "less emphasis on competition and grades" in favor of collaborative activities in "an atmosphere of friendliness and mutual support."

So children have been playing games in which everyone gets a trophy and going to schools where failure is rarely an option. Many school districts embrace a policy of "social promotion," allowing more students to go all the way through high school without mastering basic reading and arithmetic. The rationale for not flunking students and making them repeat a grade is to avoid damaging their self-esteem. But is this really what happens? The psychologist Herbert Marsh tested it by tracking large numbers of German secondary-school students for years, paying special attention to the ones who were held back to repeat a grade. It turned out their self-esteem didn't suffer that year. Instead, it rose along with their grades, and the improvement was maintained for years thereafter. Similar benefits showed up in a study of elementary-school students in Florida.

No doubt it's initially embarrassing to repeat the grade year with younger classmates, but then you adjust to them, and the material is definitely easier the second time around. So you get higher grades, feel more confident, and are better prepared for the next grade's material than if you'd been socially promoted. An unmerited promotion won't make you feel better, and it certainly won't help you learn.

Nor will an unmerited grade on your report card, but that's becoming the norm now that schools are reluctant to penalize students. Many schools have eliminated the F grade; some have eliminated grades altogether. And the ones who do give grades have gotten much more lenient. High-school grades have been rising in recent decades—the average is now a B—and it's not because students are getting smarter. Their higher grades have not been accompanied by a corresponding rise in scores on standardized tests. This overall grade inflation wouldn't matter so much if schools still ranked students against their peers, but most high schools have done away with class rank, too.

These lenient policies have had a relatively small effect on students from affluent and well-educated families, because they and their parents—and their tutors—have been aiming at the longer-term goal of getting into an elite college. They work hard because they know they'll face a penalty for not scoring well on the SAT and other standardized exams. Other students, though, are being hurt.

In 1980 half of college students were male, but today women outnumber men by nearly 3 to 2 because so many males are floundering academically. Teachers lament that boys waste time mastering video games instead of schoolwork, but instead of simply blaming the boys, they should consider the games' appeal. Players learn by competing for points and higher rankings. Instead of being shielded from failure to protect their self-esteem, they're repeatedly killed and forced to start over. The penalties enable them to learn from their mistakes and eventually earn success by outscoring other players. If school offered them the same incentives, they'd learn there, too.

Boys don't mature as quickly as girls do, so they're slower to develop self-control, and that can be especially difficult for boys in homes where there's just a single (frequently overburdened) parent to guide and discipline them. Researchers have repeatedly found that a single-parent home has more negative consequences for boys than for girls. If discipline is lax in school, too, they're more liable to struggle academically. Traditional educators relied on clear rules and penalties to keep boys on track, emphasizing competition, not collaboration, because they knew it was one of the best ways to motivate boys.

Some American educators have tried similar tactics in helping both boys and girls in low-income minority communities. As schools were embracing the everybody-gets-a-trophy philosophy, these reformers became alarmed by the worsening gap in achievement separating black and Hispanic students from white and Asian students. The gap in scores on reading and mathematics tests, as measured nationally by the U.S. Department of Education, had been narrowing in the 1970s and 1980s but then widened in the 1990s. By the end of that decade, the average Hispanic twelfth grader was scoring about the same as a white or Asian eighth grader, and the average African American twelfth grader was scoring lower. Why were so many minority students four years behind? Academics and school administrators offered various excuses for the gap, blaming it on social, political, and economic factors beyond the control of principals and teachers.

But a few principals refused to duck responsibility. They launched a movement that took its name from a 2000 book, *No Excuses,* by Samuel Casey Carter, a fellow at the Heritage Foundation, and a subsequent book, *No Excuses: Closing the Racial Gap in Learning,* by Abigail Thernstrom and Stephan Thernstrom of the Manhattan Institute. The no-excuses schools have revived traditional practices, emphasizing discipline, competition, and penalties. Instead of stressing collaborative projects, they regularly test each student's individual competence. Instead of encouraging "mutually supportive" group discussions where students are called on only when they raise their hands, the teachers challenge students by "cold-calling" them to answer a question. The schools set clear standards for both students and teachers, with immediate consequences for those who fall short.

At Success Academy, a network of forty-seven no-excuses charter schools concentrated in New York City's poorest neighborhoods, the students must wear uniforms, sit up straight during class, and walk quietly in orderly lines between classrooms. Students who break the rules or fail to try hard enough are promptly corrected, and if they persist they face further penalties, like being sent off to sit in a "calm-down chair." The schools are quick to summon parents to deal with a problem, and students are suspended more often than in the regular public-school system.

Each student is assigned a series of individualized "growth goals," like raising his or her reading skill from level F to level G within two weeks. As an added incentive, the students are assigned to one of two schoolwide groups, the Blue Nation or the Orange Nation, which compete against each other by holding pep rallies and collecting prizes for outdoing the other in progress toward their goals. When the students take overall proficiency exams, their individual scores are posted on a classroom chart, highest to lowest, with a "red zone" at the bottom indicating which students are failing. The top scorers get prizes; so do students who make the most progress and try hardest. Students with perfect attendance records get a "Zero Hero" award, and those who've been most diligent get to enjoy an "effort party," while those who've slacked off do remedial work at an "effort academy."

The students' progress is monitored not just by the teacher but also by the principal and by managers overseeing the schools. They can look directly at students' tests, essays, exercises, and other work to see how students are doing—and what kind of guidance they're getting. Eva Moskowitz, the former New York City councilwoman who founded Success Academy, says one of her first shocks at entering the education world was the low expectations for students. She originally planned to buy a conventional curriculum, but she ended up creating her own because the ones available demanded so little of students. She also found that teachers expected too little.

"When I started looking at the students' essays," she says, "I discovered that the most common piece of feedback from the teacher was, 'Good job.' Some teachers were giving almost every student an

A. We've changed that. Very few of our kids get As. It's hard to get a B. There are a lot of Cs." And if students don't get passing marks, there's no social promotion. They repeat that grade.

By monitoring students' progress each week or month, the principal and the managers at headquarters can quickly see which classes and which schools are lagging—and therefore which teachers and principals need immediate coaching and other support. The supervisors frequently sit in on classes, and when the teacher's technique is faulty they're not shy about correcting the teacher in front of the students. Sometimes they'll take over the class themselves for a few minutes to demonstrate a better way. If the extra help ultimately doesn't improve performance, then the teacher might be demoted to a tutor or an assistant teacher, and the principal might be replaced. Unlike the staff at most public schools, the teachers and principals at Success Academy don't have tenure. They face penalties for failure, including being fired.

The results of this stick strategy have been stunning—and a rebuke to the educators who despaired of closing the racial and income gaps in achievement. The Success Academy schools are publicly funded and open to anyone, with acceptance determined by lottery, so they're educating a representative sample of the students in their neighborhoods. Three-quarters of the students are poor enough to qualify for a subsidized lunch, and more than 90 percent are students of color, mainly African American. In New York City's regular public schools, the state proficiency exams are passed by fewer than half of all students, and by fewer than a third of students of color, but at Success Academy the tests are passed by 95 percent of the students. Those scores put it in the top 1 percent of all the schools in New York State. Its forty-seven schools score higher than every public-school district in the state, including wealthy suburbs that are predominantly white and Asian. The schools have become so popular in their neighborhoods that there are more than five applicants for each available seat in the lottery.

The no-excuses approach has also proved popular at charter schools in other cities—including Boston, Washington, and Chicago—and it's been validated in careful studies. Because charter schools admit students by lottery, they've provided researchers with a natural randomized experiment using the lottery losers as a control group. By tracking the lottery winners and comparing them with the losing applicants who remain in regular schools, researchers have repeatedly found that students at urban charter schools make faster progress than those in nearby public schools. And when researchers have taken a close look at which charter schools do best, they've found that most of the success is occurring at the schools using the no-excuses approach. A 2017 meta-analysis of studies of charter schools in five states and Washington, D.C., concluded that students in no-excuses schools make nearly twice as much progress annually as a regular public-school student, and that over the course of four or five years this progress is enough to erase the racial achievement gap.

Unfortunately, all this heartwarming data hasn't persuaded most educators to rethink their preference for carrots over sticks. Under pressure from federal and state laws, some regular public schools have adopted formal "zero-tolerance" policies toward certain transgressions—chiefly involving weapons and drugs—but these policies don't affect the way students are taught, and their bureaucratic rigidity has led to many instances of students in public schools being punished absurdly, like being suspended for having a penknife in the glove compartment of their car. For penalties to be effective, they must be imposed fairly, and that means giving teachers and principals the authority to make difficult decisions—and then be held accountable.

Good discipline takes work and can be painful for everyone,

which is why so many adults resist imposing penalties. No-excuses schools have faced resistance from school administrators as well as from teachers' unions opposed to penalties for teachers whose students perform poorly. The opponents have criticized Success Academy for suspending so many students and claimed it was a tactic to force out poor performers, but in fact Success Academy retains a higher percentage of its students than do the regular public schools, and the suspensions are obviously not preventing the students from learning—quite the reverse. The critics have also seized on isolated incidents to argue that there's too much pressure being put on students and teachers. After a frustrated teacher at Success Academy was surreptitiously recorded angrily criticizing a student's math mistakes and ripping up her paper, there was a torrent of criticism from progressive educators and journalists—but not from the parents at the school. They promptly rallied to the teacher's defense because they could see how much their children were benefiting from her strict standards.

The education establishment's preference for carrots over sticks is even more pronounced on college campuses, where the everybody-gets-a-trophy philosophy has become the norm. In the early 1960s, when students typically spent forty hours a week attending classes and studying, the most common grade was a C. Today they spend only twenty-seven hours, and the most common grade is an A, which is awarded nearly half the time. Again, it's not because students have gotten smarter. When researchers use external measures—tests of critical thinking and other skills administered to students at different stages of college—they find that students are learning much less today, and at a much slower pace, than they did in the past.

One rigorous study in 2011 found that nearly half of students made no significant progress in their skills at critical thinking, complex reasoning, and writing during their first two years of college, and that a third of students showed none even by their senior year. A follow-up study confirmed that those students' lack of progress hurt them after graduation: They were the ones who had the most difficulty holding good jobs. But rather than being discouraged—or galled at how much they'd wasted on tuition—the students generally expressed satisfaction with their education. They genuinely believed they'd learned a lot in college. Considering all the high grades they'd gotten, who could blame them?

Students won't learn unless someone points out their mistakes, and nothing will drive home the lesson more forcefully than a penalty. We're not suggesting a full restoration of Victorian practices—it's a very good thing that beatings are no longer routine—but we do advocate less carrot and more stick. The learning process may be more pleasurable for everyone by relying on rewards, but it will go more slowly, and the children will end up worse off. If you want your children to behave in restaurants, you're better off punishing each tantrum rather than bribing them to be polite. Denying them dessert or imposing a time-out will have more lasting impact than giving them an extra treat. It's fine to reward your children for good report cards, like paying them a cash bonus for every A, but you should also deduct something from their allowance if the grades show they've been shirking.

Some schools have tried combating grade inflation by limiting the number of As, but these efforts have faltered because of widespread resistance from students and professors (who fear students taking revenge with bad online reviews). Students argue, reasonably, that as long as all the other schools are inflating grades, they'll be at a competitive disadvantage if their school deflates grades. A more practical solution is for all schools to adopt a reform called "truth in grading" or "honest transcripts." Instead of just listing a student's grades, the transcript would also report the average grades in each course, and compute an overall comparison of the student with his peers. Besides giving graduate schools and job recruiters an idea of the students' relative ability, it would also compensate for the disparities among different disciplines. (Students in the humanities average nearly a half grade higher than science majors.)

So far only a few schools have adopted this reform, but we hope there'll be more pressure on other colleges from students, parents, graduate schools, recruiters, and public officials (like the legislators in Texas pushing to require honest transcripts at the state's public universities). In the meantime, we salute the efforts of professors like Harvey Mansfield, who has been conducting his own campaign against grade inflation. He teaches political philosophy at Harvard, where the median grade is an A minus. Mansfield gives each student two grades: one official grade for the transcript, typically an A, and another one, typically lower, that reflects their real performance. Although it's not as much of a penalty as we'd like, it's definitely better than yet another A.

A Carrot in Stick's Clothing

Not everybody gets a trophy in the adult world, yet in some ways it's not so different from school. Managers much prefer the carrot to the stick. The standard incentive for workers is a year-end bonus, not a penalty, ostensibly because it's a better way to motivate employees. But consider what happened when economists tested that principle in a small city near Chicago.

They offered two kinds of incentives to teachers in Chicago Heights, a school district where the students came mainly from low-income families and scored low on achievement tests relative to the rest of the state. A bonus pool was set up to provide merit pay averaging $4,000 per teacher (8 percent of the typical salary), with the exact payment to each teacher based on how much her students' math scores improved during the year. One group of elementary- and middle-school teachers was promised merit pay in the usual way, as a year-end bonus, but it didn't seem to motivate them. Their students' math scores didn't improve significantly.

Other teachers were given an advance payment at the start of the year—$4,000 apiece, the amount of the average bonus—and signed a contract agreeing to pay back any unearned portion at the end of the year. Their financial incentive was the same as for the first group, but in their case it was framed as a penalty instead of a reward, and that made all the difference. Their students' math scores rose significantly, apparently because the teachers really didn't want to give back any money at the end of the year.

In light of this evidence, should companies switch from the traditional year-end bonuses to a system of advance payments? It ought to improve productivity, at least in the short term, because employees would work harder to avoid a penalty. But it would also create new problems. How would a company force an employee to pay back money? It could require him to sign a contract in advance, but that means lawyers and complications. And what if he doesn't have the money at year's end? It could dock his future pay, but what if that motivates him to quit his job? And even if he does have the cash to pay the year-end penalty, what manager wants to play Scrooge during the holiday season?

But there are ways to reframe bonuses that exploit the power of bad, like the strategy tested by John List of the University of Chicago, one of the economists who did the earlier study with teachers' bonuses in Chicago Heights. The experiment took place at the Wanlida factory in Nanjing, China, where workers assembled products like digital photo frames. They were offered an extra 80 yuan per week (then worth $12, more than 20 percent of their salary) if they increased the number of units they produced. Some of the workers were told that the extra 80 yuan would be added as a bonus at the end of the week only if they met their goal. Other workers were told their regular salary was being increased by 80 yuan, but that the amount would be deducted from their paycheck any week that they didn't meet the new goal. The two offers sounded different but were effectively identical: If you meet the new goal this week, you'll get an additional 80 yuan in your next paycheck.

Both offers inspired the workers to produce more units, but once again the framing mattered. When it was presented as a penalty—a deduction from the salary—it inspired greater productivity than when it was presented as a bonus. This effect persisted week after week, providing the workers with extra pay while also improving the company's bottom line. Increasing workers' productivity was cheaper than hiring new employees, so the company's marginal production costs declined by 7 percent. Unlike the Chicago Heights school system, the electronics factory was able to use the power of bad without playing Scrooge and making workers resentful. It didn't need to force workers to pay back money, because the mere prospect of a penalty was enough.

That's one of the advantages of penalties: They're so powerful that you often don't have to use them. Rewards have to be doled out continually, but the mere threat of a penalty can make a lasting impact. In fact, one large threat can work better than many little penalties, as a food company discovered after an embarrassing case of sabotage.

The Potato-Chip Awakening

In 1969 Frito-Lay hired the puppeteer Jim Henson to make commercials for a new brand of thick potato chips called Munchos. Henson demonstrated their appeal by introducing a puppet, named Arnold, who was purple, shaggy, and gluttonous. In between ecstatic guttural grunts of "Muncho," he would gobble an entire bag of the chips, including the bag. It was an inspired commercial—the puppet would later become the Cookie Monster on *Sesame Street*—that featured a memorable slogan: "There's more to a Muncho!"

But eventually some customers discovered another meaning to the slogan. Frito-Lay started getting letters from customers who'd been enjoying their Munchos until they reached into the bag and pulled out a chip bearing a message scrawled with a black felt-tip pen: "FUCK YOU." At this point the customers tended to lose their appetite. Yes, there was more to a Muncho, but this was too much.

Frito-Lay's executives were not amused, either. After apologizing and explaining that the message did not accurately convey the company's feelings toward its customers, they traced all the obscene chips back to one plant near Dallas. Workers there had been removing freshly made chips, writing the obscenity, and putting the chips back on a conveyor belt into the packaging area. Frito-Lay's director of training and development, Dick Grote, was dispatched to the plant with instructions to "shape those knuckleheads up."

Grote, however, concluded the workers were not the only knuckleheads at the company. Frito-Lay, like many firms, had been using a bureaucratic system called "progressive discipline" to deal with employees who missed work or did something wrong on the job, like violating a safety rule. A worker's first offense would inspire a formal "verbal warning" and a notation in his file. After that was issued, there was no way for the worker to wipe his record clean. Once he was labeled a troublemaker, the next violation would automatically earn him a written warning. The supervisor would summon the worker to a meeting, hand him a preprinted form with the reprimand, and ask him to sign it. The worker, reeling from the criticism and feeling he'd been convicted without a trial, typically refused to sign it, whereupon the supervisor put another black mark in the file and sent him back to his job in a fury.

Before long there'd typically be another violation, and the worker would automatically be given a three-day suspension without pay, straining his family's weekly budget. He'd return to work angrier than ever, making the supervisor more determined than ever to get rid of him by spotting one more infraction. During the nine months prior to Grote's arrival, 58 of the 210 employees had been fired, creating a toxic atmosphere at the plant. Grote didn't see how this system of penalties would ever get them to put away their felt-tip pens. The company's idea of "progressive discipline" seemed to have degenerated into that old joke of a management edict: *The floggings will continue until morale improves.*

Grote spoke to the supervisors and found that they, too, disliked the system. They didn't enjoy hostile confrontations with workers who were frequently their friends and neighbors. Often their spouses socialized and their children played Little League Baseball together. Not wishing to strain the friendship with automatic penalties, a supervisor often ignored a worker's violations until he finally couldn't take it any longer, at which point he was so fed up that all he wanted to do was fire the worker. So he would start the formal process and go through each penalty as quickly as possible—which made the worker justifiably enraged at the unfairness of it.

Grote proposed scrapping the system and eliminating the petty penalties. Instead, the first time a worker did something wrong, there would be a meeting with the manager called "Reminder 1." This would consist of the manager noting the mistake, explaining what the company expected, and coaching the worker on how to fulfill his responsibilities. It was the technique that we recommended in the last chapter: Point out what's wrong, but then pivot to a longer discussion of how to make things right. If the worker fulfilled expectations for the next six months, the violation would be expunged from his record. If he didn't, there would be another coaching session called

"Reminder 2," after which the worker would have to put in a full year of unblemished performance in order to wipe the slate clean.

If he screwed up again during that year, he would be sent home for a day—but it wasn't presented to him as a punishment. It was called the "Decision Making Leave." He would be paid as usual, because he was supposed to be using that day to contemplate his future at Frito-Lay and one very large future threat.

"By paying for that day off, the idea was to convey how seriously the company took the situation," Grote recalls. "The supervisor would tell the worker, 'We've talked about this problem before, and you agreed to take responsibility for it. We want you to take this day to decide whether you really want to work here or whether you should head for greener pastures. If you decide to leave, no hard feelings. But if you decide to come back—and we hope you do—you need to realize that if another problem comes up, you will be terminated.'"

Grote's proposal was greeted skeptically by executives at the company. Eliminating the small penalties seemed dangerously lax, and the paid day off sounded downright strange. Without the automatic penalties, would supervisors lose their leverage over the workers? Would good workers resent a slacker being rewarded with a vacation day? Might some workers purposely get in trouble to get a day off?

But those fears proved groundless once the system was instituted. Just as Grote had hoped, it eased hostility by making the workers feel responsible for their actions instead of feeling angry at their bosses. In his interviews with workers, Grote found that the competent ones didn't mind a problem employee getting a paid day off. They didn't care how the discipline was handled as long as supervisors dealt quickly with the issue, and the supervisors started doing so because the confrontations were no longer as hostile. "The supervisors were more willing to intervene early," Grote says, "because the new system took the black hat off them and replaced it with a white hat."

And even though workers no longer faced the immediate automatic penalties, the threat of ultimately losing a job was enough of a deterrent. During the first year of the new system, the number of dismissals declined by two-thirds, from fifty-eight to nineteen. The following year only two workers were fired.

And Frito-Lay's customers stopped pulling obscene Munchos out of their bags.

Since then, organizations around the world have adopted Grote's system. This "discipline without punishment," as he calls it, exploits the long-term power of bad while avoiding short-term costs to workers or managers. Just as the supervisors at Frito-Lay dreaded being the bad guys, no one except a sadist enjoys administering punishment. And sometimes the punishment makes it harder for everyone to get the job done. A threat can be more practical than a penalty, and Grote's system isn't the only way to make one. Another method, called "discipline in abeyance," is to impose a penalty but postpone it. A worker might be given a two-week unpaid suspension but told it will not take effect until six months later—and that the penalty will be waived if he keeps a clean record until then. Instead of being angry and vengeful, the worker focuses on averting a future loss.

Ideally, the worker is also feeling something else: guilt, which is an underrated emotion. Although guilt is commonly thought of as a personal neurosis—an internal curse variously blamed on Catholic nuns, Jewish mothers, and Asian parents—it serves an essential social purpose. Most other negative emotions, like anger or anxiety or depression, tend to make people less productive, but guilt can be quite useful, as Baumeister, June Tangney, Francis Flynn, and other psychologists have repeatedly demonstrated.

Psychologists carefully distinguish guilt from shame, which doesn't have the same social benefits. To feel shame is to think, "I'm a bad person," and since you can't change your essential self, you tend to withdraw and hide or lash out in anger. In contrast, to feel guilt is to think, "I did a bad thing," and that's fixable. Guilt motivates people to improve relationships with their partners and friends by confessing, apologizing, making amends, and reaffirming their commitment. It also motivates them at their jobs. Experiments in the lab and surveys of workers have shown that the more prone people are to guilt, the harder they work at tasks and the more committed they feel to their companies. The results support the management philosophy of one of Tierney's editors at the *New York Times:* "Control the guilt and you control the reporter."

That may sound cynically manipulative, and it would be if you were making workers—or friends or students or children—feel guilty when they'd done nothing wrong. But we all make mistakes, and we all need to learn how to do better. The quickest way to help anyone improve is to use the negativity effect properly. Once the workers at the Munchos plant took responsibility for their actions, once their anger was transformed into guilt, they didn't need to be penalized immediately. A single threat in the future was enough. Grote's approach seemed revolutionary to modern executives, but in fact there was nothing new about it. George Whitefield and Jonathan Edwards knew all about inducing guilt, and they didn't bother with immediate penalties, either. When it comes to threats in the distant future, there is nothing quite like hell.

CHAPTER 6

Business 101

Yes, We Have No Bad Apples

Before she went into academia, Eliza Byington made an accidental discovery about the power of bad in the workplace. Just as negativity can be highly contagious in intimate relationships, it can also spread quickly in larger groups, often without anyone noticing. No one was aware of it at the video-production company on the West Coast when Byington took a job there.

It was an office with cubicles in a big open space, a floor plan meant to encourage collaboration, but there wasn't much collaboration going on. The sales reps who worked in the cubicles rarely chatted; the managers who had offices around the outside wall kept their doors closed. Most people went out to lunch alone or ate at their desks instead of using the communal break room. Byington considered it a fairly somber, unfriendly place until one of the sales reps developed a heart condition and started working from home three days a week.

This guy was one of the more social people in the office. He liked to chat with colleagues. But he had a caustic side. When he talked about dealing with his clients, he'd mock them and reenact phone calls to show how stupid and tiresome they were. At staff meetings, he would tease people for making little mistakes and roll his eyes when a less experienced colleague was talking. If he was unhappy with something, he'd complain loudly. Nonetheless, he was good at his job, and Byington had never considered him a problem—she'd never given him much thought at all—until he started working at home.

On the days he wasn't there, managers started leaving their doors open. Sales reps began hanging out at each other's cubicles to swap stories and strategies, and at meetings they pitched more ideas for improving the company. One of them brought in homemade brownies to share with the office. Another brought a radio to play classical music. More people began eating lunch in the break room, and some evenings a group would go out for drinks. It suddenly became a friendly place to work—but only on the days when the caustic rep wasn't there. Whenever he came back, the office went quiet again. Byington began discussing this weird dynamic with Will Felps, then a graduate student at the University of Washington business school, and they got so intrigued they looked for explanations in the academic literature.

They discovered a growing body of research into the impact of *bad apples,* a term borrowed from an old proverb that has gotten twisted recently in popular use. We now refer to bad apples as if they're isolated examples who aren't representative of the whole group. When an organization is hit by a scandal, its defenders complain about the unfairness of blaming everyone for the sins of a few deviants acting on their own—after all, the defenders say, even the best-run place will always have "a few bad apples." The linguist Geoff Nunberg dates this change in usage to the 1970 hit song by the Osmonds, "One Bad Apple (Don't Spoil the Whole Bunch, Girl)," an example of "bubblegum soul" that annoys music critics as much as it does traditional moralists.

The original proverb was coined in the days when most people lived on farms, realized that apples do not grow in bunches, and understood what happens inside a barrel of unrefrigerated fruit. "The rotten Apple spoils his companion," Benjamin Franklin warned, recycling a bit of wisdom that had been around since at least the fourteenth century, when *The Canterbury Tales* alluded to the proverb of one "roten appul" ruining all "the remenaunt" apples. Chaucer's point was that a bad apple does *not* remain an isolated example, and his grasp of social science turns out to be more accurate than the Osmonds'.

Spoiling the Barrel

Some of the pioneering research on the impact of social contacts was done with patients at the Duke University Medical Center by the psychologist Linda George, whose colleagues initially scoffed. This was back in the 1970s, and it seemed ridiculous to think that germs and diseases would be influenced by the conversations a patient had with family and friends. Over the years, though, the evidence became overwhelming. People who had plenty of visitors recovered more quickly and left the hospital sooner and healthier than those who stayed alone and unvisited. *Social support* became a buzzword among medical and psychological researchers. Receiving emotional and logistical help from family and friends was shown to boost the immune system, lower stress hormones, and reduce cognitive decline among the elderly. It made people both happier and healthier, and was linked to cheery outcomes ranging from fewer colds to longer life.

But then researchers began to consider other aspects of social networks. Yes, when you're struggling with a major illness or a career setback, it can be a boon to have a loving spouse and trusted friends who can cheer you up, help with money and logistics, provide advice and information, and offer pleasant companionship. But what if your spouse responds mainly by nagging you or belittling the problem? What if your friends say *I told you so* and offer unwelcome advice? What if they go out and gossip about your woes? And what about people with a grudge, whether they're rivals at work or a spouse already fed up with you? Soon another term appeared in the research literature: *social undermining*. Scientists started measuring both support and undermining in their statistical analyses. To the disappointment of many optimistic researchers (but not to the surprise of anyone who's read this far), the negativity effect was in force.

Social undermining was found to have a bigger impact than social support. A study of widows found that their happiness was affected more by conflicts in their social network than by the helpful things their friends and relatives did. Other studies of elderly men and women, including those dealing with bereavement or disability, found they were much more strongly influenced by bad social relationships than by good ones. When these people had pleasant interactions, they felt happier afterward, but they still worried just as much about the underlying problems in their lives. Meanwhile, unpleasant interactions made people more unhappy *and* more worried about their problems. (The lesson here is not that we should ignore the sick and the bereaved. Yes, they need support, but remember that they're particularly vulnerable to negativity. Pay close attention to their reactions and needs. As we learned in chapter 2, it's usually better to listen more and talk less.)

You'd think that someone who had just lost his job would especially benefit from contact with friends. After all, sudden unemployment tends to make people depressed and anxious. But when researchers tracked the social contacts of unemployed people for several months, they found that seeing friends eventually had its downside. In the first days of unemployment, the friends' support and undermining balanced out, but over time the criticism and the sarcasm began to have more impact than the encouragement, leaving the person even more anxious and depressed. When you're out of a job, your mental health suffers in proportion to how much grief you get from the people around you.

And when you're in a job, nothing will make you as miserable as working with a difficult boss. One study, titled "Social Undermining in the Workplace," meticulously categorized the ways that supervisors and colleagues in the police forces in the Republic of Slovenia intentionally undermined an officer, such as belittling his ideas, giving him the silent treatment, insulting him, hurting his feelings, putting him down when he questioned a procedure, making him feel incompetent, talking behind his back, and spreading rumors about him. These affronts took a toll on the worker—he became more likely to report headaches and other physical problems—and hurt the police force, too.

If the undermining was done by a colleague, the worker tended to find ways to actively retaliate against the organization as a whole, like stealing equipment. If the undermining was done by a supervisor, the worker preferred more passive forms of revenge, like wasting time on the job. The study found, as usual, that the positive actions by supervisors and colleagues did not compensate for the negative ones. In fact, the most frustrating of all situations was to have a supervisor who was alternately supportive and undermining. When a worker went to a boss expecting encouragement (because she'd offered it in the past) and wound up feeling demeaned, he became more likely than ever to sabotage the workplace.

More evidence came from a fast-food chain in Australia that allowed researchers to analyze both "good citizenship" and "deviant" behaviors by employees at three dozen restaurants. The good behavior, like being attentive to customers and helpful to fellow workers, earned employees higher ratings from their supervisors, but it didn't significantly speed up service to customers. It was the deviant behaviors that made the difference to profitability. When the staff on duty included bad workers—ones who showed up late, slacked off, or made fun of their colleagues—customers had to wait longer for their not-so-fast food, and more food went "unaccounted for" because the workers either wasted it or gave so many freebie meals to their friends.

Some corporations have tried screening for bad apples by making applicants take personality tests, which can be useful if the right measures are used. Research psychologists focus on what are called the Big Five dimensions of personality, which were derived from rigorous statistical analyses of how myriad traits are linked to one another. Unlike the grand theories of Freud and Jung, who started with broad conceptual systems of how the mind works and gradually sought to fill in the details, the Big Five is a bottom-up approach. It started with a million details, in the form of data about whether people who had one trait also had another. With the progress of statistics, researchers eventually saw how these little correlations formed into clusters: groups of traits that went together (and were separate from other groups of traits). After much number crunching and no small amount of arguing about how to run the statistics and interpret the findings, the research community settled on five major groups. (Those familiar with scientists' fondness for argument will not be surprised to hear of a recent movement to add a sixth cluster, having to do with moral traits.)

You can keep the Big Five straight with the word *CANOE*, which is spelled with the first letter of each dimension: *Conscientiousness, Agreeableness, Neuroticism, Openness,* and *Extraversion.* The ones most relevant to bad apples are the first three, which have been found to correlate with job performance. Conscientiousness is a measure of self-control, discipline, and the willingness to work hard. Agreeableness measures how well you get along and go along with others, as opposed to disagreeing a lot and picking fights. Neuroticism is a measure of emotional instability: how anxious and depressed you are, how easily you're irritated, how much complaining you do.

Emotionally unstable people can do inspired work on their own, but if you're assembling people for a joint project, your ideal is not a Vincent van Gogh or a Sylvia Plath. When psychologists watched four-person teams brainstorming during a laboratory experiment, they found, to no one's shock, that a team of emotionally stable people generated many more creative ideas than a team of unstable people. But the researchers did get a surprise when they mixed the types by putting a pair of unstable people together with a pair of stable people. Those mixed teams didn't do much better than the all-neurotic teams. There was a very slight improvement by adding the stable members, but not enough to be statistically significant. The researchers identified the familiar 4-to-1 ratio: Adding bad members had four times as much impact as adding good members.

Outside the laboratory, a similar pattern emerged from a study of the personalities of workers at several manufacturing companies. Researchers measured the key elements of the Big Five—Conscientiousness, Agreeableness, and Neuroticism—in the members of work teams. Then the researchers looked at each team's performance: how well the members communicated and avoided conflict, how much they liked one another, and how fairly they shared the workload. The scientists expected that performance would be best predicted by the average personality score of the team—after all, that would take into account everyone's strengths and weaknesses. But the strongest predictor of team functioning turned out to be the score of the worst person in the group. One lazy, disagreeable, emotionally unstable person was enough to sabotage the whole team, and it didn't matter if there was one particularly wonderful member of the group. The star couldn't compensate for the dud's damage.

Those studies gave Byington and Felps a better idea of what had gone wrong at her office. Working with one of Felps's professors, Terence Mitchell, they produced an overview of the problem titled "How, When, and Why Bad Apples Spoil the Barrel: Negative Group Members and Dysfunctional Groups." It included a taxonomy that identified three different types of bad apples. These were not mutually exclusive—a person could be more than one type—but the effects varied according to each type, as Felps demonstrated in an experiment with business students. He divided them into four-person teams that were assigned to find the best strategies for a particular business. One of the members was an actor trained to play different roles. On some teams he behaved naturally, but on others he played one of the three types of bad apple:

1. *The jerk.* The formal name in the scientific literature is "interpersonal deviant," which researchers translate to "jerk" when explaining this variety of bad apple. It's someone who subjects his colleagues to mockery, insults, nasty pranks, curses, crude jokes, and assorted forms of rudeness. When the actor in Felps's experiment went into jerk mode, he'd greet other people's ideas with questions like "Are you kidding me?" or "Have you actually taken a business class before?"
2. *The slacker.* This "withholder of effort" ducks responsibilities and fails to get his work done. He can slow down a project simply because there's less work being done, but he also has a more insidious impact. He sows discord in the group, because the others resent doing more than their fair share and so start to feel like suckers. In Felps's experiment, the actor in slacker mode would put his feet up on the desk and start texting on his phone instead of working with the team.
3. *The downer.* This "affectively negative individual" is beset by bad feelings. To get in the mood for the downer role in Felps's

> experiment, the actor imagined that his cat had just died. While the other team members debated ideas, he'd mope and lay his head down on his desk, which quickly affected the group because feelings are highly contagious when they're negative. Just showing people an image of an angry face can make them angry, and negative emotions tend to last longer than positive emotions. A team does better when it's enthusiastic, or at least not antagonistic, toward the job, but morale is tough to sustain when someone is continually glum, pessimistic, anxious, or angry. Abundant research has shown that people who interact with a depressed person end up feeling bad afterward.

In all three versions of Felps's experiment, the impact of the bad apples was obvious. On average, a team with a slacker or a jerk performed about 35 percent worse than an unspoiled team. The downer didn't significantly affect performance but caused the other members to tamp down their enthusiasm and withdraw emotionally. All three varieties of bad apple produced striking changes in the teams' behavior.

At the start of a session, most of the team members would be alert and enthusiastic, sitting up straight in their seats and ready to work. But once the actor went into slacker mode, the other people would start greeting suggestions with "Whatever" or "I don't care" or "Let's just put something down and get this over with." When the actor was a jerk, they'd become insulting and abrasive, and not just in response to him. They'd be jerks to the other team members, too. When the actor played the downer with his head on the desk, the others would soon start slouching, and eventually they'd lay their heads down, too. By the end of the session, it looked as if everyone's cat had died.

Fortunately, unless you're working at a truly dysfunctional place, you can expect to have mostly pleasant exchanges with your colleagues. When researchers went to a manufacturing firm in the Midwest and tracked employees' encounters with coworkers and supervisors throughout the workday, it turned out that the typical employee had three positive encounters for every negative one. But when the researchers looked at how an employee's mood fluctuated during the day, the negativity effect was confirmed yet again: The bad encounters were each so troubling that they had more overall impact than the many good ones. The worker might start off the morning fresh and cheerful, but one bad apple could quickly spoil the mood.

Of the three varieties, the jerk is the one who causes the most anguish. The others can slow down the group's work and dampen people's spirits, but someone who's hostile and insulting can leave his colleagues feeling bruised and aggrieved, especially if he's suggesting they're not even fit to be a member of the group. That triggers an ancient dread whose significance has only lately been identified by psychologists: social rejection.

"Nobody Chose You"

What are the most basic instincts that drive humans? Sex and aggression, Freud famously said, but his conclusions now seem dubious. Some people, after all, lead contented lives without seeking sexual partners or picking fights. While sexuality and aggression are important drives, psychologists and neuroscientists have recently focused on another urge that's much deeper and more basic. It was nicely summarized by Warren Jones, a psychologist at the University of Tennessee who spent decades studying loneliness.

"I have met many people who say they have no friends," he observed. "I have never met anyone who doesn't want to have friends."

Psychologists didn't pay much attention to this urge until the past couple of decades, starting with a paper by Baumeister that introduced a formal term for it: "The Need to Belong." He was hardly the first, of course, to recognize that people yearn for social connections. Baumeister and his coauthor, Mark Leary of Wake Forest, cited previous observations of this need, including John Donne's in the seventeenth century: "No man is an island." (They refrained from citing the more recent observation from Barbra Streisand about people who need people.) But while this desire was well known, Baumeister and Leary argued, psychologists had failed to appreciate its significance as a "fundamental human motivation." The need to belong should be almost as strong as the need to eat, because our ancestors on the savanna couldn't survive alone. Staying alive depended on being part of a group, so the human brain must have evolved with innate cognitive and emotional systems that craved frequent contacts with other humans who cared about them. That, at least, was the hypothesis put forward in the 1995 paper, which urged scientists to test it by systematically exploring the need to belong.

The first challenge was figuring out how to study this need. Baumeister initially had a hard time getting any of his graduate students interested—it wasn't yet considered a hot topic—but eventually he found a collaborator in Jean Twenge, a postdoctoral fellow who would later become known for her research on narcissism and anxiety in young people. They devised two techniques. One was to give people a personality test and tell some of them that their results indicated they were destined to go through life without a romantic partner or close friends. (At the end of the experiment they'd be reassured the results were bogus.) The people given this social death sentence reacted more negatively than other people who were told they were accident-prone and would go through life suffering one awful physical injury after another. Social rejection isn't just one of many varieties of bad news: It affects you more strongly and profoundly than other types, even the news that your future will be full of broken bones and emergency room visits.

The other lab technique was to re-create a miserable memory from the playground: the quintessentially awkward ritual of choosing up sides. These experiments would begin with a group of college students assembling in a large room to chat. The students were next led into individual rooms and asked to write down the names of two other people in the group with whom they'd like to be teamed for the next part of the experiment. Then one of the researchers would return with results—except that, as usual, they didn't tell the guinea pigs what was really going on.

Instead of tallying up who wanted to be with whom, the researchers randomly delivered one of two messages. Some people were told, "Everyone chose you." Others were told, "Nobody chose you." Eventually, after studying the reactions in both groups, the popular kids as well as the wallflowers, the researchers would mercifully reveal the fraud and send everyone home. This experimental technique became known as the voted-off-the-island protocol, after the method of exiling contestants on the television show *Survivor.*

Using these techniques, plus others, researchers have now conducted several hundred studies of the need to belong. They've developed a scale for measuring it and confirmed that it's a fundamental and universal motivation. Every single person of the thousands surveyed has expressed some desire for social acceptance. The good opinion of others matters even to famously solitary souls like the "North Pond Hermit," Christopher Knight, who shunned human contact for twenty-seven years while hiding in the Maine woods. When he was finally discovered by the police in 2013, he was embarrassed to show them his hideaway because he hadn't had a chance to tidy it up first. The need to belong is hardwired in the brain, and there's a clear pattern to it: that familiar negativity bias.

Social rejection is much stronger than social acceptance. People who suddenly feel popular, like the students who were told "Everyone chose you," typically react about the same as, or only slightly better than, the people in a neutral control group. The ones made to feel like wallflowers, however, react quite differently and generally worse. Their self-control suffers, as demonstrated in experiments showing them more prone to procrastinating on work, eating unhealthy food, taking risky gambles, and going on spending binges. They even score lower on tests of intelligence and short-term memory, whereas the popular students don't score any higher.

The wallflowers' negative responses would be still worse if not for another effect that took researchers by surprise. Baumeister and Twenge had expected a surge in negative emotions from the wallflowers, but the overall reaction was relatively muted. The wallflowers were definitely unhappier than the other students, because they registered fewer positive emotions, but they didn't show a marked increase in negative emotions. They seemed numb—and it turned out this wasn't merely an emotional effect. Subsequent experiments showed that social rejection caused a sudden slowdown in their heart rate. Their physical senses dulled, as demonstrated when their fingers were placed in a pincer that gradually squeezed tighter and tighter. Compared with the other students, the wallflowers were slower to feel any pain, and once the pain began they were able to endure it longer.

Their reaction to social rejection seemed similar to that of athletes who are injured but don't feel any pain until the game is over—a natural defense mechanism that presumably evolved because it enabled an injured person to escape from a crisis. Just as our ancestors needed a defense mechanism against attacks by animals and enemy clans, they needed one against social blows at home. ("Nobody chose you for hunting *or* gathering.")

In fact, the same neural pathways are involved, as several teams of researchers discovered. The psychologist Naomi Eisenberger scanned the brains of people playing a video game called Cyberball, in which you start out playing catch with two others but then are gradually excluded as they toss back and forth to each other. Other researchers scanned the brains of people reliving a much worse form of rejection: gazing at the photograph of a romantic partner who had recently dumped them. For both groups, the brain regions that lit up during feelings of emotional rejection were the same regions activated when they felt physical pain.

Moreover, their emotional pain could be treated with the same over-the-counter medication as physical pain. The psychologist Nathan DeWall found that giving people Tylenol eased the pain of social rejection, as measured both in brain scans of Cyberball players and in the real world. People who took acetaminophen every day for three weeks reported fewer hurt feelings from social rejection. A follow-up study showed that the use of marijuana, which affects some of the same neural receptors as Tylenol, also eases the pain of social rejection. It may seem bizarre that a pill for headaches can also treat heartaches, but it makes evolutionary sense. Social rejection can be as fatal as a physical threat, and the brain is efficiently using the same defense mechanism to deal with both menaces.

Given how crucial it is to maintain social connections, the most logical strategy for a wallflower would be to try harder to get along with others. But that's not how most people behave, at least not immediately after being snubbed. Rejection, like a physical attack, is such a shock that they react in one of two ways: fight or flight. The simpler of the strategies is flight—social withdrawal, as researchers call it. It was tested in the lab by first making students feel rejected and then asking them to do a favor for someone else, or donate money to charity, or play a game that required cooperation with another student. As usual, the negativity effect was in force.

Being socially accepted didn't make much difference—the popular students behaved similarly to the control group—but the wallflowers became significantly less helpful and cooperative. In one study, the experimenter would knock over a cup of pencils, spilling them onto the floor near the student. The students in the control group and the popular group quickly bent down to help pick up the pencils, but the wallflowers just sat there watching the experimenter pick up all the pencils. That churlish attitude wasn't going to improve their social standing—it would have been smarter for them to be extra nice—but it was self-protective, just as it's self-protective for real-life wallflowers to slink away by themselves and for jilted lovers to sit home alone listening to songs like "Harden My Heart" or "I'll Never Fall in Love Again."

The other common response is to fight—to get mad and try to get even. While the anger and aggression won't win any friends, they do help people's mood by distracting them from their sadness and anxiety. They start seeing the world through "blood-covered glasses," as DeWall summarizes the results of experiments. Once people experienced rejection in the lab, they began to assume the worst. They'd read hostility and aggression into words and actions that others considered ambiguous or neutral. They'd become more aggressive themselves when given the opportunity to punish others, including ones who had nothing to do with their earlier rejection.

It was remarkable to see such strong responses to a slight snub in an artificial setting. Ordinarily researchers have a tough time eliciting reactions in the lab. To observe aggressive behavior, as has been done in thousands of studies, you typically have to provoke people with harsh insults and direct affronts. They won't act aggressively against anyone unless that person has done something specific to offend them. But the social-rejection experiments are the great exception. Once people find themselves in the nobody-chose-you group, they'll lash out against an innocent person.

When researchers give them an outlet—by letting them rate someone else's competence for a job or administer a punishing jolt of noise during a game—they'll seize the chance to punish someone who had nothing to do with their earlier rejection. They'll retaliate against a stranger they'll never see again because social rejection is so painful even in the laboratory. Outside the lab, of course, rejection has much bigger consequences. A study of mass shootings at schools found that virtually all of the shooters had experienced acute and chronic rejection by peers or romantic partners. They attacked randomly, shooting schoolmates they barely knew, because their isolation gave them a me-against-the-world feeling.

Researchers have tested several ways of easing the effects of rejection, like instructing wallflowers to think about their relationship partners or arranging for them to have a friendly interaction with someone else. These small steps have an effect: The wallflowers become less hostile and aggressive. If you feel rejected in one group, you can take solace by reinforcing your bonds with another group. But while this tactic makes individuals feel better, it can create problems within an organization by creating animosity between groups. If a jerk on the sales team insults someone in marketing who then seeks solace from his team, the result may be a stronger feeling of solidarity on the marketing team, but they may feel resentful toward the whole sales team. Researchers have found that when people feel rejected as a group, they unite to retaliate—and they behave even more aggressively than individual wallflowers do. Their individual desires for revenge are amplified by another ancient instinct, the tribal solidarity against a common enemy. One bad apple doesn't merely spoil his companions; he can turn them against one another.

Saving the Barrel

When Robert Sutton read Baumeister's "Bad Is Stronger Than Good" paper, it immediately brought to mind his favorite term for bad apples. It had been uttered many years earlier during a faculty meeting at Stanford's engineering school, where Sutton taught organizational psychology. He and his colleagues were discussing how to fill an opening in their department when someone proposed a candidate who was known for his research (good) as well as his personality (bad). Another professor promptly objected.

"Listen, I don't care if that guy won the Nobel Prize," he said. "I just don't want any assholes ruining our group."

The rest of the professors agreed. They'd managed to keep their department a congenial haven from the bickering and backstabbing that pervades academia, and they resolved not to let anyone ruin it. They instituted an informal policy. From then on, when discussing candidates for the department, they would ask: Would this hire violate our no-asshole rule?

The rule worked so well that Sutton developed it into an article for the *Harvard Business Review* and a subsequent popular book, *The No Asshole Rule.* In looking for a scientific justification for the no-asshole rule, he discovered the literature on negativity bias and then focused on it in his own research.

"The 'Bad Is Stronger Than Good' paper has shaped my work in so many ways," he says. "It's always my doctoral students' favorite paper. That research, blended with my own observations of what managers do, has changed my view of what makes for effective leaders, teams, and organizations. The first order of business should be to eliminate the negative, not accentuate the positive." In his consulting work, he has helped companies prosper by eliminating the negative—but he has also found that most organizations still haven't learned how to do it.

The first step in dealing with bad apples is to identify them. If you're trying to decide whether or not someone deserves the label, Sutton suggests starting with two questions:

1. After talking to the alleged asshole, do you feel worse about yourself—oppressed, humiliated, de-energized, or belittled?
2. Does the alleged asshole aim his or her venom at people who are less powerful rather than at those people who are more powerful?

A more elaborate approach is to use the Big Five personality test. Researchers and corporate recruiters have found that it's hard for an applicant to fake his way through the test, and there's evidence that the results help predict how well someone will do on the job. The test is most effective at identifying slackers, because many studies have shown that a low score in Conscientiousness means trouble in any kind of work. The measures of Agreeableness and Neuroticism aren't as universally relevant—it depends on what kind of work is involved—but they can be valuable for identifying jerks and downers.

Often, though, these tests are not available or practical, and at best they can offer only guidance. There is no quick foolproof way of spotting bad apples during the hiring process. It takes work. The more pre-job interviews, the better. The ideal strategy is to hire people only after watching them in action, either during an internship or by asking the candidate to work on a project with employees. Even then, though, the candidate may be on her best behavior, so you won't discover until too late that you've hired a bad apple. At that point, there's a strong temptation to do nothing. Managers would rather avoid confrontations; workers feel powerless to do anything about a colleague. But the longer you delay, the more damage you'll suffer. Here are some principles for dealing with bad apples that have been tested and recommended by researchers and managers:

Protect yourself. The biggest danger from working with a bad apple is that he'll turn you into one, too. The process can be so subtle that most people don't consciously recognize what's happening, as Felps and Byington have found in their research. "Keep in mind just how contagious emotions and negative behaviors are—including your own," Byington advises. "Take steps to keep your own mood positive, and be aware of the mood in the group. Look for opportunities to show respect toward coworkers, and contribute to a positive emotional tone through your comments and body language." Minimize your dealings with the bad apple, and when a confrontation can't be avoided, give yourself a chance to recover afterward. Instead of getting angry or taking it out on someone else, listen to some music.

Rearrange the barrels. Sometimes the apple turns out not to be so bad once it's moved. When tensions develop on a team, quickly reassigning the manager may solve that problem without creating another one somewhere else. A manager who seems obnoxiously abrasive to one group may seem refreshingly blunt to another.

Be careful whom you label. Don't confuse useful criticism with obnoxiousness—you want people around who will speak frankly about problems and challenge others' ideas. And remember that an annoying person in one situation is not necessarily an eternally bad apple.

Don't expect bad apples to change on their own. They differ from good apples not only in how they behave but also in what they consider normal behavior. Social psychologists have identified a difference in outlook dubbed the "triangle hypothesis," which was developed by watching people take part in games that test their willingness to cooperate for mutual gain. The players who are inclined to cooperate tend to assume, accurately, that there are two kinds of players in this game: fellow cooperators as well as noncooperators who look out for themselves first. The cooperators start out hoping to work together with the other player, but if he turns out to be selfish they'll take revenge by becoming selfish, too. The noncooperators, by contrast, tend to operate on a different assumption: that all the other players are just as self-interested as they are. So they start out selfish and don't understand why so many other players are offended.

Isolate the bad apples. When the young Steve Jobs started working for $5 an hour at Atari, he literally smelled like a bad apple thanks to his fruit-heavy vegetarian diet and his aversion to deodorant and showers. He called older colleagues "dumb shits." His boss responded by assigning him to work the night shift, an isolation strategy that restored peace in the office. Other companies have successfully used physical isolation: setting up a bad apple in a distant office to work by himself.

A more daring isolation strategy, when there's a job requiring teamwork, is to put all the bad apples on the same team. The result can be awful fights, but sometimes they get along relatively well with one another. Someone who would ordinarily come off as a bully may seem normal when everyone else on the team is hyperaggressive. When the team is composed entirely of me-firsters who assume everyone else is just as selfish, at least they know what to expect from one another.

Intervene early, and don't be shy about it. You can try starting discreetly with advice and coaching, but don't expect miracles. You'll probably need to move quickly to warnings and penalties. Sutton urges companies to take into account the "total cost of assholes," as a Silicon Valley firm did with a star salesman who routinely raged at his colleagues and made his assistants so miserable that they quit.

When it came time for his annual bonus, he was presented with an itemized bill of what his behavior had cost the company during the year. It included 325 billable hours that had been spent by executives, the human-resources staff, and lawyers dealing with the fallout of his tantrums; $25,000 in overtime generated by his unreasonable demands; $5,000 for the anger-management counseling he was required to take; and $85,000 for the cost of recruiting and training a new assistant. The grand total, $160,000, was deducted from his bonus. Naturally, he was furious (it represented more than half his bonus), but he actually got off cheap, because that total didn't include all the emotional turmoil and lost productivity of his colleagues. If those costs had been included, he probably would have ended up owing the company money.

When evaluating a bad apple, look at the whole barrel. If you consider only the individual performance of a jerk, he can seem quite impressive. He may generate so much revenue that he seems too valuable to fire. But it can pay to eliminate even stars. At Men's Wearhouse, executives invoked their own version of the no-asshole rule when dealing with one of the top-performing salespeople in the entire chain of clothing stores. As impressive as his numbers were, he had repeatedly antagonized the other salespeople by refusing to help them with their customers—and sometimes trying to steal the customers away. After he was fired, none of the other salespeople at the store matched his numbers, but the store's overall sales rose by almost 30 percent.

Don't force the good apples to adapt to bad behavior. Some years ago, one of us attended a small conference designed to generate new ideas by bringing together experts from different fields. At first the conversation flowed freely and productively. Gradually, however, one person began to take over the conversation. He acted as if the group discussion were his own private seminar—a conversation between himself and everyone else—so that in effect this philosopher-king did half the talking, with the remainder divided among the nine other group members. When others tried to speak up, he would talk more loudly and thwart them. Everyone was frustrated, but no one wanted to confront the boor.

So the leaders of the group went for the classic approach: a new rule. For the remainder of the conference, those who wished to speak had to first raise their hands, and they would be called on in order. That allowed more people to talk, but it seriously hampered the discussion. Instead of responding immediately to someone's idea, you had to wait your turn, and by the time it came the discussion had usually veered off in another direction. And, of course, the boor always had his hand up, so he was continually interrupting anyway. What was supposed to be a productive brainstorming session ended up being a collection of disjointed comments and ideas that went nowhere.

The one productive result came later. When they drew up plans for the next meeting, they made sure not to invite the boor. This time they actually got something done as they batted ideas back and forth, and they didn't have to institute any protocol for raising their hands. They had made yet another independent discovery of the no-asshole rule, and once again it worked.

Don't hesitate to fire a jerk, but don't be a jerk about it. If the other strategies haven't worked, get rid of the bad apple as soon as possible. But no matter how much trouble he caused, be kind. It wasn't just his fault. This was a mutual mistake—you shouldn't have hired him in the first place—and he's already going to bear the brunt of the consequences. Generously offer him whatever help you can—time, advice, severance pay—to find another job that suits him. That's not merely the decent thing to do. It's also the smartest strategy. Even though he'll no longer be around, how you treat him will be duly noted by everyone left behind, and the negativity effect will be in force. One bit of cruelty to a sacked employee will be remembered much longer than anything nice you do for someone else.

CHAPTER 7

Online Perils

The Sunshine Hotel vs. the Moon Lady

The Casablanca is not the most famous hotel in New York City, and it's hardly the most luxurious, but it is the most extraordinary. It has beaten bad online. The Internet has amplified the negativity effect, giving cranks and slanderers and trolls instant access to bigger audiences than ever, but the Casablanca has figured out how to prevail in this new arena. Every night nearly five hundred hotels in New York compete to earn good reviews on TripAdvisor, and every morning the Casablanca triumphs. It gets so many 5-star ratings that it has ranked among the city's top five hotels every single day for more than a decade, typically in first place.

The Casablanca streak is all the more remarkable considering the ingrates who have stayed there. They book the smallest, cheapest room—listed as having just one narrow bed—and then post a 1-star review because it won't accommodate four people. They don't bother to open the window and then whine about the lack of fresh air. They complain that the complimentary breakfast is "not worth the money" and sneer that the free wine served at happy hour is the "$12 a bottle variety." Many are appalled to discover that there is street noise in midtown Manhattan. A 1-star reviewer complained that there were crowds of people near the Casablanca, a problem he apparently never imagined when choosing a hotel that touts its location on West Forty-Third Street "steps from Times Square." Another critic declared himself "summarily disappointed" to discover that Times Square is "crass"—a failing that he blamed, of course, on the hotel.

And then there was the Moon Lady. She checked in after midnight and informed the hotel, for the first time, that she required a room with a view, and not just any view. She needed to see the moon. Since the hotel is a six-story building surrounded by skyscrapers, this would have been a difficult request to fulfill under any circumstances, and at that moment—one A.M.—the only empty room was the one reserved for her on the fifth floor.

The clerk, who was not equipped with an astronomical chart of that night's lunar trajectory, said he couldn't guarantee the moon but hoped the room would otherwise satisfy. It did not. Upon inspecting it, she called the front desk to complain that the room had no windows at all. The clerk, knowing it had two windows with wooden blinds, volunteered to go up and raise the blinds so she could see the windows, but she refused the offer and declared the room unacceptable. The clerk offered to provide her with another room at a sister hotel, but she refused that offer, too. Instead, she took her suitcase, left the hotel, and posted a review on TripAdvisor describing the horrors of the Casablanca.

"If this is your first time in New York City and you stay in this motel, most likely you will never return to New York again," she wrote. "There was NO WINDOW. NO WINDOW in the room, I couldn't believe it. . . . The closet size room without window is like sleeping in the coffin." The 1-star review was headlined "Very Dissapointed!!! Don't Go There!!!"

In the pre-TripAdvisor era, a hotelier could happily say good-bye to a guest like her, safe in the knowledge that she would never darken his lobby again. But today guests like her never really leave, not when their 1-star reviews remain for years on TripAdvisor, frightening away thousands of potential guests. Whatever your business, unhappy customers matter more than ever in the online world.

They've gained so much power that there's a new underground industry exploiting the negativity effect. Blackmailers shake down hotel managers and restaurateurs by posting a bad review on TripAdvisor, Yelp, Google, or Facebook, sometimes accompanied by staged photos of dirt or vermin, and then offering to delete it if the bill is waived. (The blackmailers know how devastating a single accusation of food poisoning or bedbugs can be.) A business has to deal with fraudulent reviews posted by a rival—or by hired guns working for the rival—and by wildly inaccurate reviews posted by customers. The problem has gotten so bad that there's a cottage industry of counternegativity experts like Adryenn Ashley, a consultant in San Francisco who helps businesses deal with bad reviews and collects horror stories at a website called yelp-sucks.com.

"I get a tearful phone call at least once a week from someone who says that Yelp is putting them out of business," she says. "They feel helpless, victimized, alone, worthless, like somehow it's their fault. They're blaming themselves for not understanding how Yelp works and not knowing how to make a bad review go away." The perils of Yelp and TripAdvisor are specific to business, but negative online comments can provoke those same feelings of despair in anyone who ventures on social media—which is coming to mean everyone.

The Casablanca could not make the Moon Lady's review go away. It's still up on the TripAdvisor website, readily visible to the many travelers who go straight to the 1-star reviews when they're considering a hotel. But the Casablanca is thriving anyway, because it has learned how to deal with the online power of bad.

The Fault in the Stars

Suppose, in checking the online reviews for an apartment complex called Maple Grove Towers, you come across this review by someone named Pat:

> The neighborhood is nice and the people in this building are kind and friendly. The rent is affordable. I love this place and would recommend this place to anyone.

How much weight would you give to that review? When researchers tested its impact by posting the review on several websites created for the experiment, they found that the answer depended on where the review appeared. The review seemed less credible when it was posted on a personal site called *Pat's Blog* than when it appeared at ApartmentReview.com, which was billed as "Apartment Reviews by Real People."

But the readers had a different way of evaluating Pat's review when it was negative:

> The neighborhood is bad and the people in this building are mean and unfriendly. The rent is not affordable. I hate this place and I would not recommend this place to anyone.

On the face of it, a couple of things seem odd in this version of the review. Is *everyone* in the building really mean and unfriendly—or is

Pat just too unpleasant to get along with anyone? If the rent is "not affordable," why did Pat move in, and how is Pat still living there? Did Pat get evicted for some reason and now wants to take revenge on the landlord? Does this review have more to do with Pat's personal issues than with problems at Maple Grove Towers?

If you thought about those questions, it would be logical to be more suspicious of this review than the positive one, and you'd therefore pay more attention to where the review was posted. But in the experiment, people paid *less* attention. They didn't care at all. They considered the complaints just as credible whether the review was on *Pat's Blog* or at ApartmentReview.com. No matter who Pat might be, the negativity effect swayed their judgment of Maple Grove Towers. They wouldn't wish that hellhole on anyone.

Other studies confirm this online negativity bias: People planning a vacation spend more time studying negative reviews and are more influenced by them than by positive reviews, and people are swayed even by negative reviews that don't identify any specific problem. An analysis of sales trends at Amazon and Barnes & Noble revealed the familiar negativity effect for both fiction and nonfiction books: A 1-star review did more to hurt sales than a 5-star review did to boost sales. Angry people like Pat or the Moon Lady do so much damage that business researchers place them in a special category of customers—"terrorists"—a term coined well before the 1-star reviews on TripAdvisor and other online sites. Studies of old-fashioned word of mouth showed that dissatisfied customers were more likely than satisfied ones to discuss their experience—and also to tell more people about it.

Some of the pioneering work was done in the 1980s at Xerox, which each month asked a sample of forty thousand customers to rate their satisfaction on a scale of 1 to 5, with 3 being neutral. The initial goal was to convert all customers to a 4 ("somewhat satisfied") or a 5 ("very satisfied"), but then the company discovered that 4 wasn't nearly good enough to guarantee repeat business. Those "somewhat satisfied" customers weren't much more loyal than the neutral or dissatisfied customers. Realizing that any niggling discontent could kill a sale, Xerox changed its goal to getting all its customers up to 5.

It was the same story with car buyers, airline passengers, hospital patients, telephone-company customers, and computer shoppers. After analyzing tens of thousands of their ratings in the 1990s, researchers at the Harvard Business School concluded that the only loyal customers are the ones who give you a 5 rating—and most of them aren't going to spread the word about your business. Only a minority of the 5s are enthusiastic enough to be "apostles," as they were categorized by the researchers. You can't expect apostolic work from the rest of the 5s or any of the other customers. Most of them fit into the researchers' category of "mercenaries" who will instantly switch to another company if they can get a better deal. Some are "hostages" who buy your product only because they have no choice—and will gladly escape if there's a new option. Worst of all are the "terrorists," who are not only dissatisfied but also determined to spread the bad word.

Which customers should you pay most attention to? The Harvard researchers, Thomas O. Jones and W. Earl Sasser Jr., calculated that the returns would be meager from trying to woo neutral or slightly dissatisfied customers. It would be more lucrative to concentrate on turning 4s into 5s, and ideally into apostles. But the biggest payoff of all would come from focusing on the most unhappy customers. Preventing a customer from turning into a terrorist would yield a fourfold return on investment—not from that customer, but from future customers who otherwise would have been scared off.

Today the power of terrorists is vastly amplified. The Canadian musician Dave Carroll became both a folk hero and a case study in business schools after United Airlines refused to reimburse him for a guitar smashed by its baggage handlers. He wrote a song, "United

Breaks Guitars," and posted a series of videos that have been seen more than twenty million times on YouTube. Most unhappy customers don't have Carroll's flair for complaining, but with social media it doesn't take much talent to be a terrorist. When McDonald's wanted to increase its "Twitter presence," it invited customers to tweet about their experiences and promoted a hashtag, #mcdstories. Plenty of customers tweeted praise, but so did critics with tales of food poisoning and accusations that eating there would lead to diabetes ("McDialysis, I'm lovin' it"). Within two hours the promotion was abruptly ended by McDonald's social-media director, who explained that "#mcdstories did not go as planned." He pointed out that fewer than 2 percent of the seventy-two thousand tweets that day about McDonald's were negative, but those were the ones that kept getting quoted in articles and blogs, and retweeted along with another hashtag: #McFail.

Fortunately for businesses, the web also makes it easier for satisfied customers to be heard. The average review rating has risen in the past decade to between 4.0 and 4.5 stars out of 5, which is partly because businesses like the Casablanca Hotel have learned how to turn more customers into apostles. While happy customers aren't as passionate as unhappy customers, they're motivated by the urge for "public self-enhancement," a term that psychologists use because it sounds less obnoxious than "showing off."

Suppose you pride yourself on your knowledge of food and local restaurants. After a mediocre dinner at an overpriced French restaurant with slow service, you could vent your frustration with a withering review on Google or Yelp or OpenTable. But if you're such an authority on local cuisine, why did you choose that restaurant and waste so much of your money? A nasty review doesn't necessarily reflect well on you, either, so you may not want to post it. You'd sound savvier by rhapsodizing about some little gem of a Turkish restaurant you discovered. This self-enhancement strategy would lead people to post mainly positive reviews about their own experiences, and that's just what researchers observed during experiments.

When people discussed something they'd paid for themselves, they mostly reported on their favorite meals and successful shopping expeditions. But when they talked about *other* people's experiences, they preferred to tell about awful restaurants or gadgets that didn't work. The results of the experiments were reported in a paper titled "On Braggarts and Gossips," which is a nice summary of how word of mouth operates both online and offline. People like to brag about their triumphs and gossip about others' failures.

It's not that the braggarts are lying. Most customers really are satisfied—that's how businesses stay in business. A single bad experience still has a disproportionate impact—an unhappy Amazon book buyer is more likely than a happy customer to post a review—but there are so many more happy customers that the typical book gets 4 stars. The early reviews of a book tend to be especially positive, and these raves not only sell books but also encourage other fans to post their own praise.

Once a bad review is posted, though, the negativity effect can start to skew the discussion. Others become reluctant to praise something even if they liked it themselves. In one experiment, after people gave their opinions of a brief clay-animation film, some of them were shown a review of it on a website and asked to post their own review. When they saw a positive review, it didn't affect their judgment—they posted the same rating as did people who hadn't seen any review at all. But when people saw a negative review, they tempered their own enthusiasm for the film and posted a lower rating. They didn't actually change their private opinion of the film, but they became less willing to express it publicly. No one wants to come off as a naïf who lacks the discernment to spot flaws.

This anxiety causes many satisfied customers to avoid posting anything once a bad review appears, and researchers have repeatedly found that the ratings tend to drop over time for a book on Amazon and for other online products. As soon as a product gets one or two scathing reviews, other customers lose interest in posting reviews, and as the online discussion slows down, it comes to be dominated by a minority of activists—the kind of customers who like to distinguish themselves from the crowd by posting lots of negative reviews of different products.

In fact, some of them will pose as customers for things they haven't bought. In one national survey, a fifth of Americans confessed to posting reviews for products or services they hadn't used. A study comparing reviews and purchases at an online clothing retailer found that 5 percent of the reviewers hadn't bought the clothing—and that these busybodies were twice as likely as the real customers to give a 1-star review. The same pattern was observed in book reviews at Amazon: The "verified" customers known to have bought the book on Amazon gave higher ratings than the other reviewers, some of whom were probably panning books they hadn't read. It's not easy to spot these frauds, but researchers have helpfully identified a few signs to watch for. Fake reviews, either positive or negative, tend to be a little longer than average but give relatively few specific details about the quality of the product or the business, instead offering general opinions or extraneous information. They also tend to have more exclamation points!

Online pans get more attention partly because of our negativity bias, and partly because we're understandably suspicious of positive reviews. We know that the raves about a restaurant's cuisine or a novel's prose could be coming from friends and family—or the connection could be even more direct. When a glitch at Amazon's Canadian website briefly revealed the real names of some reviewers, it turned out—to no one's surprise—that some of the 5-star reviews of books were coming from the authors. But online fraud goes both ways. There's a black market in reviewers for hire who will either boost your business or trash your competitors. One ad on Craigslist solicited people with "established Yelp accounts with over 50 reviews" to provide "well-written reviews for a restaurant" at $25 per review. If you're less picky, you can find online offers from bulk operators who will post fake reviews for $5 apiece.

Chris Emmins, a British consultant who helps businesses combat online defamation, identified one reviewer on TripAdvisor who claimed to have stayed in fifty-one hotels in Paris in a single month—while also finding time that month to stay at dozens of hotels in Germany, Italy, and Spain. Emmins further embarrassed TripAdvisor by submitting fake reviews of his own that propelled a restaurant to seventeenth place in the London rankings, ahead of thousands of rivals, even though the restaurant had been closed for months. No one knows exactly how prevalent this sort of fraud is, but it's statistically significant, as demonstrated in an ingenious study by a team of marketing researchers led by Dina Mayzlin of the University of Southern California.

By comparing the hotel reviews on TripAdvisor and Expedia, the researchers noticed a suspicious pattern when there was a local independent hotel competing with a nearby chain hotel. In that situation, the owner of the local hotel presumably cares more about generating business through online reviews, which aren't as important to chain hotels because travelers already know what to expect from a Marriott or a Hampton Inn. The local hotelier needs to woo out-of-towners away from a known brand, so he would be tempted to arrange for his hotel to get raves and his chain competitor to get panned. And he would be especially tempted to do that on TripAdvisor, because it allows anyone to create an identity and post a review for no charge, whereas Expedia allows reviews only from customers who have actually paid (through Expedia) to stay at the hotel.

Sure enough, when the researchers did the comparisons, they

found that the local hotel would consistently have a higher share of 5-star ratings on TripAdvisor than on Expedia, while the nearby chain hotel would have more 1-star reviews on TripAdvisor than on Expedia. Somebody was trying to make the chain hotel look worse on TripAdvisor, and the obvious suspect was right next door. In general, the local fraudsters put more effort into boosting themselves than into tearing down their rivals—they posted more fake raves than fake pans, perhaps because they felt less sleazy posing as apostles rather than terrorists. But the pans still had more impact. And this could have a big effect on a hotel's bottom line, because other research has shown that a 1-star decline in ratings forces a hotel to charge about 10 percent less for its rooms.

The arithmetic becomes especially cruel for businesses near the top of the rankings, where there are so many rivals with nearly perfect ratings that it takes just one or two bad reviews to plummet in the standings (and disappear from the first web page seen by potential customers). The owners of Botto Bistro, a trattoria near San Francisco, got so frustrated with the disproportionate impact of negative reviews on Yelp that they launched their own counteroffensive. On their Facebook page and their menu, they mocked the arbitrary criticisms on Yelp and set a goal of becoming the lowest-rated restaurant. They offered a 25-percent-off coupon to anyone who posted a 1-star review of their restaurant, and their loyal customers gleefully trolled Yelp by writing 1-star reviews with absurd complaints. The olives had pits inside! The water was terrible! Sure, the pizza was delicious, but there wasn't any good Chinese food! This awful restaurant used fresh ingredients instead of canned sauce and vegetables! The staff was unable to fix our car's flat tire!

That was one way to counteract the negativity effect, and it worked. The mock negative reviews swamped the impact of the real ones, and the restaurant attracted lots of customers who got the joke and appreciated the feisty assault on Yelp. But much as we admire the spirit of Botto Bistro's owners, we don't think their strategy will work for most businesses. The Casablanca Hotel is a better role model.

The Peak-End Rule

What's the secret sauce? Adele Gutman gets asked that question a lot. She is the mastermind behind the success of the Casablanca as well as the half-dozen other boutique hotels of its parent company, the Library Hotel Collection. The hotels in New York, Toronto, and Prague all rank perennially in TripAdvisor's top ten for their cities, and the one in Budapest, the Aria, took TripAdvisor's annual award in 2017 as the number one hotel worldwide.

So Gutman has been barraged with queries from other hoteliers and invitations to give master classes at trade conferences. She can talk the business-school talk, expounding on "best practices in reputation management" and offering mantras like "Service Is Marketing," but there's one phrase she keeps coming back to: "sparkling sunshine." She says it with a smile and a fluttering of her perfectly manicured fingers to illustrate the sunshine her staff is sparkling over every guest.

"You have to double up on the good things," she says. "If you manage to connect with every single guest, you've given yourself an insurance policy against bad reviews because they're not likely to say something negative about somebody who's their friend. You have to go over the top so they forget the bad things. I never use phrases like 'meeting people's expectations' or 'satisfying customers.' I say 'sparkling sunshine,' and our staff gets exactly what I mean."

There is nothing haphazard about this sunshine. It's a system she developed after taking over the marketing of the Casablanca and its sister hotels in 2005, when they were ranked lower on TripAdvisor. She realized that they couldn't compete with low-end hotels for price or high-end hotels for luxury. They were small hotels with a sense of style—the Casablanca had a Moroccan theme taken from the Humphrey Bogart movie—but they didn't offer palatial suites or sweeping views. She also knew, though, that luxury was not the formula for getting to the top of TripAdvisor's rankings. Deluxe hotels in New York like the St. Regis and the Plaza were routinely outranked by cheaper hotels because their guests expected so much and would find something to complain about. The secret to making that crucial first web page on TripAdvisor was to avoid negative reviews.

After studying the reviews, she drew up a list of all the "contact points" between a guest and the hotel, from making the reservation to checking out, and resolved to sparkle sunshine at every point. The front desk started keeping a diary listing every request or complaint from a guest and how it was handled. Gutman focused on hiring cheery extroverts and coaching them to engage the guests whenever possible. The telephone reservation agents at the Casablanca don't just book a room; they ask why the guest is coming to New York and if there's anything special they need.

From the doorman to the front-desk clerk to the bellhop, everyone is supposed to beam—"Welcome to our hotel!"—and treat the guest's arrival as a singularly delightful treat: "Oh, this is your first time in New York? We're going to have fun with you! The favorite part of our job is helping people make the most of New York. If you want any recommendations or help, please, please, let us know." When the bellhop shows the guests to the room, he watches their reaction and reports back to the manager. If the guests seem unenthusiastic, the manager will call to make sure it's all right and offer another room if possible.

That welcome may seem like overkill, and no doubt some weary travelers would rather check in without all the fuss. But this strategy makes perfect sense to researchers who have studied how people form judgments. First impressions really do matter, and they're definitely governed by the negativity bias. Some of the clearest evidence comes from tracking reactions of people administering job interviews. When the candidate makes a good first impression, the interviewer will be swayed only slightly, and that mildly favorable impression can be quickly reversed. But if a candidate comes off badly in the first moments, he'll have to spend the rest of the interview trying to make up for it, and he'll be lucky to get back to neutral.

If the job interview ends on a sour note, the candidate had better keep looking elsewhere, because a bad last impression is even worse than a bad first impression. It's an example of what psychologists call the peak-end rule, which was demonstrated by having people immerse their hands in ice water. First they dunked their hands for sixty seconds. Then, after a break, they dunked their hands again, but this time they kept them there for ninety seconds, with the water getting slightly warmer during the final thirty seconds. Later, they were told they had to undergo one more dunking and asked which version they'd prefer.

Most people preferred the second version. Although it lasted longer and involved more overall pain, afterward it seemed less painful because it had ended with slightly warmer water. The effect was confirmed in studies of patients' reactions to a colonoscopy or the removal of kidney stones. The duration of the procedure and the total amount of pain mattered less than the combination of two other factors: the peak level of pain and the level of pain at the very end.

A less painful demonstration of the peak-end rule occurred one Halloween when researchers from Dartmouth staked out a home and carefully doled out candy to trick-or-treaters. Some of the children got a single treat, a Hershey bar, while others received first the Hershey bar and then a piece of bubble gum. When asked to rate the candy at that home, the children who got both treats were *less* satisfied than the ones given only the Hershey bar. That measly piece of gum at the end soured the experience. It may sound childish to be less happy with more candy, but adults reacted no differently to the gifts they received in another experiment. The ones who got DVDs of a good movie and a mediocre movie were less satisfied than the ones who got just the good movie. The moral for gift givers: Save the best stuff for last.

The peak-end rule is relevant to both bad and good experiences—but not equally, of course. Suppose your supervisor is about to divvy up a year-end bonus pool, and you're being compared with a fellow salesperson. She was consistent week to week in generating money for the firm, whereas you had more ups and downs, but over the course of the year you both averaged the same amount of revenue. How will your supervisor reward each of you? Variations of this scenario were tested by researchers in South Korea who expected, in accordance with the peak-end rule, that the supervisor would pay more attention to the extreme weekly highs and lows than to the annual average.

It turned out the researchers were half right. If your most extreme week was a bad one, your supervisor would indeed rate you lower than your colleague even though you'd gradually made up for it over the course of the year. But if your most extreme week was a good one, you'd end up with the same bonus as your colleague. Instead of being swayed by the spectacular figure that week, your supervisor would see that it was counterbalanced by the many weeks in which your colleague did slightly better.

The peak-end rule helps explain why reviewers on TripAdvisor will rant about an unpleasant surprise at checkout—"Beware the minibar bill!"—or fixate on the sleep they lost because of street noise. The complaints can seem ridiculously petty—"I was extremely upset to find only one tube of shampoo in the shower"—but Adele Gutman takes them all seriously.

"Traveler reviews are like a free customer focus group," she says. "Even when they're unfair, you can learn something from them." She has eliminated obvious sources of irritation at checkout time by not having minibars in the rooms and providing free bottles of water, free Wi-Fi, and free breakfast. To avoid unpleasant surprises, the website offers photo tours of each room with painstaking details on what's there (the size of the room and the bed) and what's not ("no view of the city"). It warns that front rooms facing Forty-Third Street get more street noise, and the back rooms get less light. Guests with sleeping problems are urged to take advantage of the hotel's "Escape to Serenity Program," which offers mattress toppers, an assortment of special pillows, earplugs, white-noise machines, and headbands equipped with built-in headphones to play soothing sounds.

Gutman has also created one more "contact point" with the guests by luring them into the lounge throughout the day, where complimentary snacks and coffee are offered around the clock, and there's a reception every evening with wine and cheese. The point isn't just to propitiate the guests with freebies. It gives Gutman and the staff more chances to sparkle sunshine and forestall complaints. "When you're constantly taking the guests' temperature," she says, "you can find out if there's some little thing they were too shy to ask for—something that could be the difference between a four-star and five-star review." As we've seen, listening to bad is a crucial step in overcoming it.

Thanks to all its strategies, the Casablanca has maintained a 5-star rating on TripAdvisor for over a decade. Close to 90 percent of the reviews are for 5 stars, and only 3 percent are below 4 stars. While some of Gutman's tactics are peculiar to the hotel industry—most businesses don't offer a chance to mingle with customers every day at a wine-and-cheese reception—the basic strategies can be applied in other businesses. Gutman's secret sauce is a set of techniques for overcoming the negativity bias and the peak-end rule:

1. *Focus on making a good first impression.*
2. *Look for ways to create many more good impressions. (If "sparkling sunshine" seems too corny a mantra, come up with your own.)*
3. *Anticipate and eliminate any irritant that could become a negative peak.*
4. *Keep monitoring your customers' reactions to watch for unanticipated problems.*
5. *When a complaint arises, respond quickly no matter how petty it seems.*
6. *Don't just correct something bad. Overwhelm it with good.*
7. *No matter how crazy or obnoxious the customer, end on a good note.*

Most transactions end with the customer paying the bill, hardly the best of notes, but smart businesses can find ways to blunt the pain, as restaurants do when they provide a free dessert or present the check along with some complimentary chocolates. Netflix prospered in its early days by eliminating the late fees that infuriated customers returning movies to their local video store. Retailers like L.L.Bean, Lands' End, IKEA, and Nordstrom placate unhappy customers by letting them return items long after the normal thirty-day grace period. Rental-car companies and hotels have learned to combat sticker shock by warning customers in advance of all the taxes and fees that will be added to the bill.

But some businesses remain stubbornly oblivious to the peak-end rule. Why do so many shopping expeditions end with a long line at the checkout counter, and so many airline flights end with a half-hour wait at baggage claim? Why do so many online newspaper articles end with a correction informing the reader of some trivial mistake made in an earlier version? In the print era, running corrections was the only way to set the record straight once the paper had come off the presses, but today any error can be instantly rectified online so that future readers won't see it. Why should they care if a past version misspelled someone's name or gave the wrong date for the Battle of Constantinople? Corrections are warranted for genuinely serious errors, like a false accusation against someone, but otherwise they're of interest only to a few sticklers who could be mollified by running all the corrections on one web page and sparing ordinary readers. As it is, newspapers are guaranteeing that an otherwise satisfactory experience ends on a gratuitously bad note: *Even if you liked this article, we want you to know that we have a really incompetent staff, and we're so ashamed of them that we'll waste your time telling you about a mistake you never saw.*

The "end" part of the peak-end rule is especially tough in the online world, because a customer who leaves angry is much likelier to post a review than a happy one. To offset that negativity bias, Gutman sends all guests an email after their departure urging them to post a review, and she subscribes to a service that instantly streams the latest reviews from TripAdvisor and dozens of other sites. Some businesses make a point of posting a response to every review, but the Casablanca usually doesn't if the review is positive—a wise strategy, according to Chris Anderson, a professor at Cornell's School of Hotel Administration who has analyzed the effects of responses on TripAdvisor. He finds that hotels improve their online ratings and boost their revenues by posting responses, but only if they don't overdo it. Potential customers get annoyed—and start shopping for another hotel—if they have to keep scrolling through thank-you notes for 5-star raves. He advises responding to no more than 40 percent of the reviews, and to concentrate on the negative ones, just as Gutman does.

As soon as a review for 3 or fewer stars appears, Gutman or the hotel manager will post an online response and write the guest directly. Often the guest will be sufficiently mollified to delete or revise the review. Even when the critic leaves the review intact, the response ensures that their anger isn't the last thing that readers will see. This particular story will end on a good note.

"We were devastated to hear you did not enjoy your one night stay with us," the Casablanca's manager, John Taboada, wrote to a 1-star reviewer complaining about street noise. To show how devastated he was, he explained that he'd discussed the situation with the staff and went through all the efforts they'd made to accommodate this guest, like trying to find him a quieter room and offering him a white-noise machine and other sleeping aids (which, as Taboada tactfully noted, the guest had declined).

Taboada's responses can run to five hundred or a thousand words, much longer than the critical reviews—and again, that's the point. It's a business version of the strategy we discussed earlier in dealing with a romantic relationship: Listen to the other person, take their negative reaction seriously, and offset it with lots more positivity. By explaining all the details of the situation, all the efforts of his staff, and his new plans for making sure this will never ever happen again, he's overwhelming the bad and getting the last word. The reader goes away thinking not about the critic's complaint but about the hotel's devotion to pleasing guests: *They're nice even to the worst cranks.*

The Casablanca has thrived by sparkling sunshine even on guests like the Moon Lady. After ranting about the windowless coffin that was her room, she ended her review with a vow: "I will be reporting this incident anywhere I can. This is not right. No tourist should ever go through such experience anywhere." The Casablanca could have fired back with an account of her bizarre behavior, but instead it posted a conciliatory note directing readers to a photo of the room with its two windows. Instead of pointing out the obvious, that this reviewer was insane, it offered soothing advice to future guests: "If you have any special needs, such as a room with a view of the moon, it is best to tell us in advance so that we can let you know upfront if we are unable to meet your request." The Moon Lady never showed up again at the Casablanca, but plenty of other travelers did.

CHAPTER 8

The Pollyanna Principle

Our Natural Weapon Against Bad

The Pollyanna principle is a lot better than it sounds. It's a powerful psychological effect, based on solid research showing we have innate defenses against the negativity effect—some unconscious, others that can be consciously deployed. It just happens to be named for an irritating character from a really cloying book.

The heroine of *Pollyanna,* the 1913 novel by Eleanor H. Porter, is a saintly orphan who remains impossibly cheerful through a series of gratuitous cruelties, random tragedies, preposterous coincidences, and cringe-making dialogue with equally implausible characters. Somehow, though, none of that prevented it from being an instant bestseller. The novel, followed quickly by a sequel, became a Broadway hit that launched the career of the teenage Helen Hayes.

The film rights were sold for the then-astronomical sum of $115,000 to Mary Pickford, the star known as "America's Sweetheart." She produced the movie and cast herself, at age twenty-seven, as the twelve-year-old Pollyanna. The script was assigned to another of Hollywood's leading talents, Frances Marion. Both women were appalled by the novel's maudlin scenes and sentiment, as Marion later noted in her memoir. "We proceeded with the dull routine of making a picture we both thought nauseating," she recalled. "I hated writing it, Mary hated playing it."

Yet somehow, once again, Pollyanna prevailed. The 1920 movie was a smash. Pollyanna became a defining role in Pickford's career, just as the 1960 Disney version was for Hayley Mills. The Pollyanna franchise has now survived more than a century, selling millions of books and inspiring more films, television series, board games, comic books, and a dozen sequels by other authors.

Pollyanna endures because she offers something valuable: a psychological strategy for overcoming the negativity effect. What researchers call the Pollyanna principle is a genuine insight no matter how annoying she sounds when describing it.

She reveals it early in the novel, shortly after arriving at the home of her aunt in a small Vermont town. The aunt, a nasty piece of work who lives alone in a mansion, greets Pollyanna coldly and leads her upstairs, where the girl marvels at the lush carpets and furnishings in the unoccupied bedrooms. But then Pollyanna finds she's been relegated to a tiny room in the attic. She quickly sees an advantage to the bare walls and lack of curtains: nothing to distract from the view out the window! Pollyanna explains to her aunt's servant that she's playing a game her late father taught her back at their mission outpost in the West, starting on the day some donated goods arrived. Pollyanna had asked for a doll, but the only item in the shipment for a child was a pair of small crutches. She was disappointed until her father taught her to always look for some reason to rejoice: Be glad you don't *need* the crutches!

Pollyanna teaches this Glad Game to her new neighbors, and it promptly brightens the lives of everyone in town. Even the mean aunt learns to smile and is duly rewarded with a wonderful husband. Pollyanna suffers a brief crisis of faith after she is hit by a car and her legs are paralyzed, but then her spirits rally, and at the end of the novel she miraculously walks again. You can see why the movie appalled Pickford and other Hollywood luminaries like the director D. W. Griffith. He called it "the most immoral story ever produced on screen" because of its "false philosophy of gilded bunkum."

But it wasn't bunkum. Bad art, yes. But as science, the Glad Game's not bad at all.

Happy Talk

If you have a pen or a keyboard handy, here's a twenty-second exercise to try before reading the next paragraph: Make a quick list of words associated with emotions.

Unless you've already fallen under the influence of Pollyanna, your list probably contains more words for bad feelings than good feelings. Studies in Chicago, Mexico City, and a half-dozen European countries showed that people consistently list more words like *angry* or *afraid* than words like *happy.* Scholars who pored through dictionaries in a variety of languages reached a similar conclusion: There are lots more words for bad emotions than for good emotions. These negativity biases in language make intuitive sense, given what we know about the power of bad. People pay more attention to negative emotions, so there are more ways to describe them. But which words do people use more often? Here the answer is not so intuitive.

Researchers have been studying the question since the 1930s, when the federal government was casting about for ways to put people back to work during the Great Depression. Some unemployed scholars were hired to count the words in books and other publications. It turned out that *good* appeared five times as often as *bad,* and *better* appeared five times as often as *worse.* The words *love* and *sweet* were used seven times as often as *hate* and *sour.* Mentions of happiness outnumbered those of unhappiness by a whopping 15-to-1 margin. Subsequent researchers did small-scale studies in a dozen other languages and found a similar positive bias. After surveying this evidence, the psychologists Jerry Boucher and Charles E. Osgood published a paper in 1969, "The Pollyanna Hypothesis," proposing that people everywhere tend to use positive words "more frequently, diversely and facilely" than negative words.

"Pollyannaism is hypothesized to be a human universal," they wrote, but it wasn't until the age of Big Data that the hypothesis could be properly tested. In 2012 a team of applied mathematicians at the University of Vermont published an analysis of the 5,000 most commonly used English words in some 300,000 songs, 2 million *New York Times* articles, 3 million books, and 800 million tweets. Despite our biased sensitivity to the negative, Pollyannaism reigned in all the media. Even in the songs, which you'd expect to be heavy on lovers' laments, the ratio of positive to negative words was about 2 to 1. The positivity ratio rose to nearly 3 to 1 in tweets and nearly 4 to 1 in books and newspaper articles.

The Vermont mathematicians then went after still bigger data globally. Working with researchers at the Mitre Corporation, they developed a computer algorithm called a hedonometer to measure the emotional content of texts in ten languages: books, news articles, song lyrics, film and TV scripts, websites, and social-media posts. Pollyannaism again reigned in every medium and in every language. Even grim novels like *Moby Dick* and *Crime and Punishment* had more positive than negative words overall (although the concluding sections were definitely downers).

Now that the Pollyanna hypothesis has been confirmed—and upgraded to the Pollyanna principle—the next question is why it exists.

One possible explanation is simply that good is more common than bad, so we have more good things to talk about. Even newspapers, those great purveyors of ill tidings, turned up quite positive on the hedonometer analysis because the depressing stories from around the world were outweighed by other sections of the paper reporting on athletic victories, artistic achievements, philanthropic bequests, groundbreaking ceremonies, weddings, and awards. As we noted earlier, studies of diaries show that people typically have three good days for every bad day, so there are more good things in their lives than bad things.

But the diary research has also shown that the bad days have a bigger impact: If today is bad, it increases the chances that tomorrow will be bad, too, but a good day doesn't carry over to the next. Other researchers have found that when people are asked about good and bad events in their lives, they'll spend more time thinking about the bad ones and will tell longer stories about them. As news editors know, people also spend more time reading horror stories than feel-good features. So the mere frequency of good events doesn't fully explain why we keep using positive words. Given how much extra attention we pay to bad, there seems to be something else prompting us to sound upbeat, as Boucher and Osgood speculated when they proposed the Pollyanna hypothesis in 1969.

"Why do most people most of the time in most places around the world talk about the good things in life?" they asked. "The answer surely goes beyond psycholinguistics *per se* and into the nature of human social structures and the conditions under which these structures can be maintained. It is hard to imagine human groups whose members persistently look for and talk about the ugly things in life and in their neighbors long remaining together."

There are some obvious social benefits to accentuating the positive. People like being flattered, which requires saying nice things, and they find upbeat people appealing. By digitally altering facial expressions in photographs, researchers have found that adding a smile to a face causes onlookers to judge the person as more attractive, generous, healthy, and agreeable. Positive language confers some of the same advantages. In chapter 4, we noted that a critic faces the choice between seeming "brilliant but cruel" with a negative review or "plodding but kind" with a positive review. Coming off as kind is usually the best social strategy, at least for people who aren't professional critics.

Sounding positive makes you more likable and can also lift the spirits of your audience, whether you're talking to them in person or online. By tracking popularity on Twitter, researchers found that people gained more followers by tweeting positively rather than negatively. Another study showed that people on Twitter tended to turn positive or negative in their posting depending on what tweets they saw, but they were more eager to adopt upbeat themes. While negative tweets attracted more immediate attention—they'd be retweeted more rapidly—over the long haul it was the positive tweets that proved more popular and spread more widely. They were retweeted twice as often as negative ones, and they were "favorited" five times as often. The Twitter wars that attract so much attention in the press are not the norm. Some people love to fight online—chiefly the political class, a group we'll discuss later—but most prefer to accentuate the positive.

Of course, each individual negative tweet or post has more impact than a positive one, which is why it's still easy enough to get upset on some sites. Of the major platforms, the one least likely to bring you down is YouTube, at least according to the British Royal Society for Public Health. It compared social-media platforms by surveying the heaviest users, teenagers and young adults. They reported lots of positive feelings and few negative ones after spending time on YouTube, which is understandable considering what they were watching and sharing with one another: very little politics or other news, lots of music videos and TV shows, plus other popular content like how-to videos, comic bits, inspirational stories, and homemade videos (including all those adorable animals). On the whole, YouTube left them feeling less anxious, depressed, or lonely.

The other social-media platforms produced some benefits, too, leaving the users with greater feelings of community and emotional support from their friends. But there were also some negatives reported from the peer pressure on those platforms. The photos of other people's wonderful lives and marvelous physiques on Facebook, Snapchat, and Instagram engendered that mixture of envy and dread known as FOMO—fear of missing out—as well as anxiety over body image. The platform generating the most anxiety over body image was Instagram, for an obvious reason: Instagram has become the go-to marketing platform for celebrities and models showing off their expensively sculpted bodies, and for personal trainers posting "fitspiration" photos of people who seem to live in the gym. Some researchers have reported that these images are making people dissatisfied with their own bodies, but it's important to put these results in perspective.

We need to be wary of the "Fredric Wertham effect," named after the psychiatrist in New York whose shoddy research inspired a moral panic over youths' media preferences in the 1950s. His book, *Seduction of the Innocent,* was condensed in *Reader's Digest* under the headline: "Comic Books—Blueprints for Delinquency." Eventually, after congressional hearings and dozens of local bans, the comic-book scare was debunked, but researchers and journalists have kept repeating his mistake. They've rushed to conclude that youths are being corrupted by new scourges like television, rock music, rap, video games—and now social media, which is being blamed for the supposed afflictions of "Facebook depression" and "Instagram envy."

The press has eagerly reported studies linking social-media use to loneliness, anxiety, and depression, but many of the alarming conclusions are based on dubious correlations and problematic experiments. Meanwhile, careful studies have produced more reassuring (and less publicized) results. Researchers have reported that social-media users actually have more close relationships than other Internet users, that many reap psychological benefits, and that more social networking doesn't lead to more depression or other psychological and behavioral problems. Some users do feel bad, but they tend to be people who were already troubled by insecurities.

There's no doubt that young people struggle with peer pressure and hostility and ostracism online, but they've always struggled in the real world, too. After reviewing social-media research, the psychologist Christopher Ferguson concludes that the online harms have been hyped—and that young people are still mainly affected by their offline peers and interactions. His meta-analysis of studies on body dissatisfaction finds that the images on social media have no significant impact on men's satisfaction with their bodies, and only a small impact on some women: those who were already quite worried about the issue.

The lesson is not to avoid social media but rather to use it wisely. Researchers who have analyzed online social norms and users' feelings find that most people consider it a duty to avoid insulting others while presenting themselves "positively but honestly," and that it is indeed a winning strategy. People who post more positive messages are considered more attractive, get more social support in return, and consequently feel happier, whereas those who post downbeat messages get less encouragement and can end up feeling worse.

So it pays to be Pollyannaish in what you post, whom you follow, and which sites you visit. If you're feeling insecure, spend less time looking at elaborately staged and retouched photos of celebrities and models, and more time looking at snapshots of your friends and family. (And if some of your friends keep posting FOMO-inducing photos of their vacations, be glad you have the freedom to unfollow them.) To stay positive, follow people who spread optimism and civility rather than outrage or vitriol—and there are plenty of these happy campers on social media.

The old mass-media dictum "If it bleeds, it leads" doesn't govern social media. There's a big difference between what people read and what they share. When researchers at the University of Pennsylvania analyzed the *New York Times'* most emailed list for six months, they found that negative articles were less likely to be shared than positive ones. People couldn't help reading about scandals and shootings, which were often at the top of the most viewed list, but they preferred sharing articles that were cheery ("Wide-Eyed New Arrivals Falling in Love with the City") or that inspired awe, like a new theory about the structure of the universe.

This disparity between what we heed and what we discuss was explored by neuroscientists in an experiment measuring social buzz. By scanning the brains of people who were being exposed to new ideas, the researchers could see which ideas they found most exciting: the ones that lit up the brain regions associated with encoding and retrieving memories. But afterward, those weren't the ideas that generated the most conversation. The most buzzworthy ideas were ones that lit up a different region of the brain, the one associated with social cognition—thoughts of other people. In deciding which ideas to pass on, the people focused not on the most personally exciting but on what would be most appealing to others.

People do discuss bad news, of course, but even then they lapse into Pollyannaism. The hedonometer team's analysis of more than one hundred billion tweets worldwide showed that Twitter users become more negative in response to terrorist attacks and other bad news, but then there's a rebound, and even on the worst days tweeters use more positive words than negative words.

"While there are terrible stories in the news and awful threads on Twitter, we tend not to go on about them," says Peter Sheridan

Dodds, a leader of the hedonometer team at the University of Vermont. "Language is our great social technology, and we use it to help us get through hard times. Language can encode behaviors and thoughts we're unaware of, and it seems a bias to positivity is one of them."

This positivity bias goes deeper than the language we use. It's not just a happy front that we put on for others. Psychologists have also found positivity biases in ways that we remember the past and see the present—internal manifestations of the Pollyanna principle. We use nostalgia for the past as a way to feel better today. Bad will always affect us more strongly than good, but we've developed conscious and unconscious versions of the Glad Game to blunt its impact.

The Pollyanna Within

For most of the last century, psychologists have been coming up with reasons not to be happy—and not to expect things to improve, either. Psychology textbooks devoted twice as much space to unpleasant rather than pleasant emotions, and there was a similar negativity bias in research journals. Psychologists emphasized the bad impacts of events: the lifelong neuroses stemming from childhood memories, the enduring effects of post-traumatic stress, the debilitating effects of aging, the fear of death that supposedly haunted us all.

At best, we were stuck running in place on a "hedonic treadmill," because not even good events could permanently cheer us up, as illustrated in a famous study in 1978 of lottery winners. At first, winning the lottery brings joy—but it wears off. When asked how they felt a year later, it turned out the jackpot winners weren't happier than their neighbors or any more optimistic about the future. In fact, they weren't any happier than another group of people who'd been in accidents that left them paralyzed, a finding that seemed to show we're stuck on the treadmill no matter what happens to us.

The hedonic treadmill became a staple in psychology textbooks, and the lottery study became a favorite of journalists. They'd cite it as evidence for the popular belief in the "curse of the lottery," which was trotted out when a big winner wound up divorced, depressed, destitute, or dead. But that one study was far from conclusive. It involved only twenty-two lottery winners in Illinois, and it didn't actually measure how the jackpot affected their happiness. It merely recorded their feelings at one point in time, typically within a year of winning the jackpot, and compared them with those of some of their neighbors. The researchers themselves acknowledged these limitations and urged further studies tracking winners' feelings before and after they won the lottery.

That research has finally been done, thanks to a couple of studies of British lottery winners. Some of the winners' psychological well-being dipped slightly in the year immediately following the jackpot, and some of them tended to drink and smoke a little more in that first year as they adjusted to their good fortune. But those effects soon eased, and after two years the winners were significantly better off psychologically than they'd been before hitting the jackpot. So much for the curse of the lottery. Money really did buy happiness—and that's good news for those of us who haven't won the lottery, because it means we're not necessarily stuck on a hedonic treadmill. We just need to find other ways to cheer up, and there's increasing evidence that we can do so by tapping our inner Pollyanna.

Researchers are revising the traditional view of psychological trauma, which emerged after World War I soldiers were diagnosed with the novel condition of "shell shock." Later called "Post-Vietnam syndrome," it was eventually known by the broader term *post-traumatic stress disorder* (PTSD). The disorder was quite real—and one more manifestation of the negativity effect. Some bad events, unlike good events, would affect people for decades or for life. When a bad event had permanent effects, like an accident that left someone unable to use their arms or legs, it could permanently reduce a person's level of happiness. (This finding was another blow to the hedonic-treadmill theory.)

But then, starting in the 1990s, psychologists noticed something else, too. While many people—at least half the population, by some estimates—endured a traumatic event at some point in their lives, most didn't show symptoms of post-traumatic stress. Four out of five trauma victims did *not* suffer from PTSD afterward. And in the long run, they typically emerged stronger. Instead of being permanently scarred, they underwent *post-traumatic growth* (PTG), a term introduced by the psychologists Richard Tedeschi and Lawrence Calhoun. PTG isn't nearly as well known as PTSD (good is never as newsworthy as bad), but it's far more common. Studies have found that more than 60 percent (sometimes 90 percent) of trauma victims undergo post-traumatic growth, including ones who initially showed symptoms of PTSD.

This growth isn't a result of the trauma, which in itself is inherently bad and produces harmful consequences. Even the author of *Pollyanna* couldn't imagine a silver lining for the paralysis that befalls her heroine. "If I can't walk, how am I ever going to be glad for—ANYTHING?" Pollyanna asks in despair. But before long, as the townspeople flock to her aunt's home to tell her how she's transformed their lives, Pollyanna is back to playing her game. Reflecting on the good she's done, she beams and says, "I can be glad I've HAD my legs, anyway." As nauseating as her reactions seemed to Mary Pickford, they jibe with the checklist that psychologists now use to chart post-traumatic growth: increased appreciation of life, deeper relationships with others, new perspectives and priorities, greater personal strength. The growth comes not from the trauma but from the way that people respond to it to become kinder, stronger, and more mindful of the joys in life.

They suppress their negativity bias with an array of defenses that are available to anyone. Even though a bad event triggers a stronger immediate reaction than a good event, negative emotions typically fade more quickly than positive emotions. This "fading affect bias" isn't universal—in depressed people, the bad emotions linger longer—but it's been repeatedly observed in experiments tracking people's feelings. First they come into the lab to describe how they feel about recent events, and later they return to recall those same events. By then all their feelings have diminished, but the negative ones have faded more than the positive ones, especially among people who have spent the most time discussing the bad events with others. As we noted in chapter 1, the negativity effect distorts people's judgment most strongly when they feel personally threatened. They're not as susceptible to overreaction when something bad happens to someone else. So the more you talk with others about your problems, the more perspective you can gain to ease your anxieties.

Whether or not you get outside help, there are mechanisms within the brain to reduce the sting of bad. Earlier we discussed a couple of these: the "positive illusions" of women with breast cancer (who form unrealistically optimistic but nonetheless useful expectations) and the tendency of happily married spouses to suppress the brain's critical faculties when evaluating their partners. People also create positive illusions of the past. Researchers like to say that we look back through rose-colored glasses, which is why older people seem to have fewer unhappy memories than young people do. This is illogical—obviously more bad things have happened to someone who has lived longer—but it's been observed over and over. For instance, when the parents of infants and toddlers are asked if they've ever regretted having children, many will quickly say yes. But when the question is put to parents with grown children, they're more likely to say no, they never regretted it for a minute.

What happened to the memories of the three A.M. diaper changings and the toddler tantrums? Those bad moments are "Forgotten but Not Gone," to quote the title of a study that tested people's memory of information about themselves that was either positive or negative. The people could recognize the unflattering stuff just as well as the praise when they were prompted with clues. But if they were asked to recall things off the top of their head, without getting any hints, they couldn't dredge up as many bad things. The bad stuff was still in their memories, but it was tucked away in less accessible places. It could be prompted to resurface—and other studies have shown that involuntary memories, the kind that seem to pop into our heads out of nowhere, are more likely to be negative. But when we make a conscious effort to think about the past, we put on those rose-colored glasses.

Sports fans have much more vivid memories of a championship season than of the many years the team didn't win. The bad memories fade with time, but they relive the victories over and over in their minds and in their conversations. Lottery players have the same tendency, as a colleague of ours noticed by spending some time with them in line to buy tickets. They loved telling her about their past wins. When she politely asked about past losses, they quickly changed the subject back to their wins.

Frequent gamblers have to be particularly adept at the Glad Game, because the negativity effect makes their habit so difficult to maintain. After all, losses have more impact than gains, and most gamblers lose more than they win. What keeps them going? An answer emerged from an experiment in which experienced sports bettors gambled on the next Sunday's games in the National Football League. After the games were played, the players settled their bets and were asked by the researchers to explain what had gone right and wrong. Then, three weeks later, they were asked to recall the games they'd bet on, and if there'd been any particularly important plays in each game. You might expect them to dwell on the games they picked correctly, if only to feel better and impress their listener with their acumen, but it turned out they engaged in a more elaborate form of Pollyannaism.

They focused their attention on "near wins"—the bets they figured they should have won because they'd correctly picked the better team only to see it lose because of a fluke, like a bad call by the referee or an unlucky fumble. They spent more time talking about those losing bets, and the fluke plays in the games, than about their correct predictions. It was their self-serving method of recalculating their won-lost record. If their team won, they automatically considered it proof of their betting savvy and didn't need to think much about it anymore, even if their team played badly and won thanks to one of those fluke plays. But if their team lost a close game, then they focused on the fluke in order to convince themselves that they'd been right all along. And if they were given a chance to bet on a rematch between those teams, they'd pick the same team again even though it had lost the first time. Instead of learning from their mistakes—the reason that the negativity bias evolved—they'd override the bias and double down.

When it comes to gambling, not even the original Pollyanna would recommend playing the Glad Game. Rose-colored glasses can become dangerous blinders. But when there's no money at stake and no crucial lesson to learn, the glasses can be quite useful. We noted in chapter 3 that people's social anxieties can be eased by training them to focus on friendly faces instead of hostile faces when they go into a room. This technique seems to come naturally to people as they age. Experiments tracking eye movements have shown that older people are more likely than younger people to look toward smiling faces and away from frowning or scowling faces. When shown someone's face with a surprised expression, younger people tend to assume the person has been startled by something bad, whereas older people are likelier to interpret it as a look of joyful excitement.

This "positivity effect," as researchers call it, has been observed in older people in dozens of studies testing what kinds of words, images, and stories command attention and stick in people's memories. Compared with younger adults, people over sixty are likelier to recall photographs of smiling babies than ones of wrecked cars. Brain scans show that the emotional regions of their brain respond less to negative images and more to positive ones. When hearing themselves criticized, older people get less angry than younger people do. When deciding which car to buy, which doctor or hospital to choose, or which gift to take home from a laboratory experiment, they pay more attention than younger people do to the upside of each choice, and less attention to the downside. After they make a choice, they're more satisfied with it than young people are.

Their inner Pollyannaism showed up clearly in a brain-scanning experiment in Germany involving a video game with a series of seven closed boxes. All the boxes contained gold except for one containing an obnoxious little devil. The players would open one box after another, keeping any gold they found inside, and they could go on as long as they wanted, but they'd lose it all if they opened the box with the devil inside. So they typically opened a few boxes but then played it safe by quitting while they were ahead, whereupon the researchers would show them where the devil was lurking—and how much more money they could have made by playing a little longer.

What effect did this missed opportunity have? The researchers analyzed three groups of players: older people who had been diagnosed with depression, mentally healthy older people, and mentally healthy young people. It turned out that the younger people, whose average age was twenty-five, reacted similarly to the group of depressed older people, whose average age was sixty-six. The thought of that lost money triggered heightened activity in the brain regions associated with regret, and it also affected the way they played the next round of the game. The healthy younger people and the depressed older ones would go on to take bigger—and stupider—risks.

Only the healthy older people were able to avoid regret, as demonstrated both in the brain scans and by their subsequent play. By focusing on what they'd won instead of lamenting what might have been, they not only remained happier but also went on to win more gold by playing the game more astutely. The German researchers, led by the neuroscientist Stefanie Brassen, concluded that "disengagement from regret reflects a critical resilience factor for emotional health in older age."

Of course, older people also have more to regret than younger people do. They have more aches and pains to deal with every day. They've endured more disappointments. One of us once heard an old New Yorker sitting on a bench in the middle of Broadway who summed up his life in one sad sentence: "I made *every* mistake you can make in the garment business." That line seemed ominous at the time—was this the miserable gloom awaiting us all in old age?—but it's a relief now to see evidence that the old man wasn't typical. In his regret, he was actually more like a twenty-five-year-old than other people his age. Most people overcome disappointments to find more joy as they get older, and we know this not just from the many laboratory experiments demonstrating the positivity effect.

The most compelling evidence for Pollyannaism in old age—and yet another blow to the hedonic-treadmill theory—comes from surveys around the world asking people to rate their happiness and their satisfaction with life. Over and over, researchers have found that happiness declines in middle age and then rebounds. It typically bottoms out around age fifty and then keeps rising for decades, so that people in their sixties are happier than they were in their twenties, and they remain happier in their seventies and eighties. This U curve of happiness, as it's known because of its shape, has been observed in dozens of countries, and it's been confirmed by tracking people's happiness over the course of decades.

It seems most pronounced in developed countries, where people tend to get wealthier as they age, but it's not due merely to having more money. Even when researchers control for income and other factors (like employment status and education), they find that people's life satisfaction rises in their old age. Further evidence of this pattern comes from trends in European countries of antidepressant use, which peaks in the late forties and then goes down in old age. People can turn morose near the end of life, when they're struggling with serious illness and disability, but until then they generally find that the golden years live up to their name.

This pattern has even been observed in our fellow apes. It showed up in an international study of more than five hundred chimpanzees and orangutans in zoos and wildlife sanctuaries. When their human keepers were asked how positive the animals' mood generally was, and how much pleasure they seemed to take in social interactions, it turned out that the older apes (the ones in their forties and fifties) got higher ratings than the middle-aged ones. The older apes weren't as prone to negativity bias, presumably for some of the same evolutionary reasons as humans. Older animals don't need to learn as much as younger animals do, so they don't need to pay as much attention to their mistakes.

They may also reap health benefits in the process, because negative emotions and stress weaken the immune system. Experimenters have found that older people with stronger immune systems also tend to be better at recalling positive images. This finding bolsters the theory that the positivity effect is an evolutionary adaptation: As the body's immune system weakens with age, there are automatic processes to turn us mellower. Older men produce less testosterone, which makes them less aggressive and more compassionate, while older women produce less estrogen, which makes them less anxious and more confident. These hormonal shifts, as well as other changes in the aging brain, could help explain why older people are better at regulating their negative emotions.

But their Pollyannaism is also due to their own conscious choices. The positivity effect diminishes in studies of old people with cognitive problems, like Alzheimer's, and also in experiments that make it difficult to concentrate. When experimenters ask people to keep track of sounds at the same time they're looking at images, older people aren't any better than younger people at recalling happy pictures of puppies and babies.

Their positivity doesn't all come automatically. It takes mental effort, whether they're looking at photos in a laboratory or maintaining a sunny outlook in their daily lives despite the travails of aging. They're deliberately playing the Glad Game, and some of their strategies can work at any age.

Glad Games

At the age of forty, Constantine Sedikides found himself beset several times a week by a strange new feeling. He had just moved from the University of North Carolina to England, to teach social psychology at the University of Southampton, and he kept getting hit by waves of nostalgia for his old home. Several times a week, he'd suddenly start thinking about old friends in Chapel Hill. He'd remember their lamb barbecues and Thanksgiving dinners, the strong taste of the local coffee, the sweet smell of the humid air on fall football weekends. The intensity of the memories surprised him. One day, at lunch with a colleague on the Southampton faculty, he mentioned how good he felt after these bursts of nostalgia.

"I don't believe a word you're saying," replied the colleague, a clinical psychologist. "You don't live your life by reliving it. That's a clinical sign of maladjustment." He peppered Sedikides with diagnostic questions. Was he depressed? Listless? Had he lost interest in sex?

Sedikides insisted all was well, but the colleague remained convinced that nostalgia meant trouble, and he had plenty of support in the psychological literature. Nostalgia had been in disrepute since the word was coined in a seventeenth-century Swiss medical dissertation to identify "a neurological disease of essentially demonic cause."

The medical student, Johannes Hofer, combined the Greek words for homecoming (*nostos*) and pain (*algos*) to describe the homesickness observed in Swiss mercenary soldiers fighting abroad. It was considered so debilitating that Swiss officers banned, under penalty of death, the playing of an Alpine milking song that supposedly triggered the disease. Military physicians theorized that the Swiss were particularly susceptible because their eardrums and brain cells had been damaged in their youth by the incessant sounds of clanging cowbells, but doctors soon identified the dread symptoms in other countries, too. Nostalgia was classified as a form of "melancholia" and identified during the nineteenth and twentieth centuries as an "immigrant psychosis," a "regressive manifestation," and a "mentally repressive compulsive disorder."

But Sedikides didn't feel melancholy or repressed, much less psychotic. He was known on the social-psychology circuit for his infectious laugh and playfulness. Born in Greece, he'd dropped the first letter of his original last name, Tsedikides, so that it would read the same backward or forward. (He considered completing the palindromic effect by changing his first name to Bob, but decided that he just didn't look like someone named Bob Sedikides.) He'd enjoyed his time in North Carolina, where he'd studied the way people filter information to enhance their image of themselves, but he wasn't longing to return there. The bursts of nostalgia left him feeling better about his life and more energized to tackle his work in Southampton.

After hearing his colleague's skepticism at lunch that day in 1999, he began studying nostalgia, and ended up pioneering a field that has grown to include dozens of social scientists.

They've coined a verb, *nostalgizing,* to distinguish this sentimental longing for the past from mere reminiscing, and induced it in laboratory experiments by playing hit songs from people's youth. They've used a questionnaire to measure levels of nostalgia around the world. Most people report nostalgizing at least once a week, and many do it three or four times. Whether they live in England or China, people tend to do it more often on cold days, and experiments have shown that nostalgizing in a cool room actually makes them feel warmer. They also do it more often when something bad happens, or when they're feeling down—and it boosts their spirits, as repeatedly demonstrated in the lab and in people's daily lives.

Nostalgia isn't what it used to be, not after two decades of this research. Far from being a pathology, it has been found to counteract loneliness and anxiety. Couples who nostalgize more frequently have better relationships. Inducing nostalgia inspires people to write more creatively and act more generously (a finding that won't surprise the college fund-raisers who stoke these feelings with alumni reunions). Nostalgia motivates us to finish tasks, overcome adversity, and strive toward goals. When people are prodded by researchers to engage in a few minutes of nostalgia before starting the workday, they're better at coping with stresses on the job (like a boss who's rude or unfair). By broadening our perspective, nostalgia gives us the sense that our lives have more meaning. Sedikides likes to quote a character from *Mr. Sammler's Planet,* the Saul Bellow novel, who explains his fondness for looking backward: "Everybody needs his memories. They keep the wolf of insignificance from the door."

Nostalgia isn't a purely positive emotion. It's bittersweet, tinged with a sense of loss, but it's ultimately rewarding. Literature's first nostalgist, Odysseus, pined every day for his family back in Ithaca, but the memories galvanized him to overcome obstacles on his way home. When researchers systematically analyzed the stories in the magazine *Nostalgia* (yes, there's a magazine devoted to it), they found that the positive emotions greatly outnumbered the negative, and a similar happy imbalance showed up in the stories collected during experiments in various countries. People everywhere have fond memories of holidays, weddings, picnics, and hikes.

"A typical nostalgic memory," Sedikides says, "will start off with a problem that you solve with help from someone else. For instance: *I go to the family reunion, and Uncle Harry comes up and says, 'Tina, have you gained weight?' Then Auntie Martha comes up to me and says, 'Oh, Tina, don't pay attention to Uncle Harry. You know how he is. Come have a drink with me.' And then we have a great time talking about the family.* The story starts badly with stupid Uncle Harry, but it ends well. Nostalgia makes you feel more connected to other people and gives you a sense of continuity with your past, so you feel happier and have a greater sense of meaning in your life."

People of all ages engage in nostalgia, even children as young as seven, who like to replay their memories of birthday parties and vacations. But the level varies with age. Nostalgia dips in middle age and then rises again in old age—the same U-shaped curve found in plotting people's happiness over their lives, and it's not a coincidence. Sedikides and his colleagues have found a direct correlation: The more time that adults spend nostalgically reflecting on their past, the more satisfied they are with the present, and the more optimistic they become about the future. As people age, they become happier by getting better at reaping the benefits of nostalgia. These results have inspired Sedikides to do more nostalgizing of his own, and he'd like to see more of it in other people of all ages.

It's one of the promising strategies to emerge since psychologists began compensating for their profession's century of negativity bias. In the past two decades, since Martin Seligman launched the positive psychology movement, they've focused less on what goes wrong and more on how people overcome it. What gives trauma victims the resilience to emerge stronger? Why do older people report feeling better even though their bodies are worse? Some researchers hypothesized that the positivity effect in old age was due to cognitive decline—the aging brain avoided negative information because it was more difficult to process—but that theory hasn't held up. Older people can still focus on the negative when they need to make important decisions, and in experiments they make those decisions as well as young people do.

They're not simpleminded Pollyannas. They don't ignore problems and bury bad feelings, but they're not overwhelmed by bad. They find benefits from setbacks, the way that trauma victims find new strengths. They're better at dealing with complex emotions, the way that nostalgists can feel happy even while they're wistfully thinking of past joys that can't be repeated. They focus not on what's been lost but on the good memories that remain. They pay less attention to the hassles of daily life and more attention to the small pleasures.

With experience comes wisdom, and with age comes a new perspective. One of the leading explanations for the positivity effect in older people is the shift in their time horizon: With less time left to live, they focus more on enjoying the present than on striving toward long-term goals. They devote more time to close friends and relatives, and less time to cocktail parties and blind dates. They avoid the power of bad by exercising "socioemotional selectivity," as the psychologist Laura Carstensen calls this explanation for the positivity effect. Younger people ordinarily focus more on the future because they have more years to live, but they, too, are capable of shifting perspective. When they're prompted to focus on the present, either during experiments or when they're reminded of the fragility of life, they show the same sort of positivity effect as older people—and gain some of the same benefits.

In *Pollyanna,* it was an adolescent who taught everyone the Glad Game, but in reality the young must learn it from their elders. They'll never be as good at it, because they need to pay more attention to the bad. Young people starting their careers and looking for romance must learn from their mistakes and broaden their horizons—they've got to endure cocktail parties and blind dates. Middle-aged people have to confront the emotional and financial challenges of raising children and caring for their aged parents. They feel more stressed because there are more stresses in their lives.

But everyone, young or middle-aged or old, can get better at the Glad Game. No one is stuck on a hedonic treadmill that limits happiness. Here are a few specific strategies drawn from the recent research into positive psychology:

Change the narrative. Wounded soldiers and accident victims experience post-traumatic growth by rewriting the story of their lives. They see the injury not as something that shattered their plans but as something that started them on a new path. This same technique can be used for any kind of bad event in life. Being fired from a job can be seen not as a failure or a career killer, but as the impetus that leads to a better career. Experiments with trauma victims and other people have shown marked benefits from devoting fifteen minutes a day to "expressive writing." This means writing about your problems and your feelings about them. The exercise forces you to confront the bad in your life and helps you move beyond it by dealing with it.

Share your good news. Of all Mark Twain's aphorisms, the one with the most empirical support is a bit of wisdom from the title character of *Pudd'nhead Wilson:* "Grief can take care of itself, but to get the full value of a joy you must have somebody to divide it with." Psychologists could quibble with the first part, because there is some benefit from sharing grief, but they've consistently found much greater benefits from sharing joy.

Their term for it is *capitalization.* The psychologists Shelly Gable and Harry Reis have studied its effects by analyzing diaries and watching people in the laboratory talk about good things that have happened to them. When someone responds enthusiastically to your good news, you feel happier and the triumph seems more significant. It helps you savor the moment—and as Sedikides has found in studies of college students and alumni, the more you savor an experience today, the more likely you are to enjoy nostalgizing about it in the future.

Rejoice (or at least fake it) when you hear someone else's good news. It takes two to capitalize on good news. If the listener just sits there quietly instead of reacting enthusiastically, nobody benefits, and the speaker may just end up feeling deflated. Researchers have shown that one of the quickest ways to improve a couple's relationship is to train them to celebrate each other's daily victories. The technique is simple enough: You listen carefully to the good news, put a smile and some energy into your response ("Wow, that's great!"), and follow it up with questions to elicit more details. That way the victory looms larger, and the victor feels not only happier but also closer to their partner and more trusting and generous. It's a win-win that confirms another literary aphorism, this one from the poet John Milton: "Good, the more communicated, more abundant grows."

List your blessings. Cultivating an "attitude of gratitude" is one of the most effective strategies identified by the positive psychology movement. It has been linked to less anxiety and depression, better health, and higher long-term satisfaction with life. When people are prodded to count their blessings, they become less aggressive when provoked and act more kindly toward others, including their romantic partners. They fall asleep more quickly at night, sleep longer, and wake up feeling more refreshed.

Older people are naturally better at it than younger people, but anyone can use the technique tested by researchers: Write a list of five things for which you're grateful. It can be a specific event of the day or something general like "wonderful friends." A daily journal is ideal, but experiments show there are benefits from doing it just once a week. Another proven strategy is the "gratitude visit": Write a three-hundred-word letter to someone who changed your life for the better, listing specific reasons, and then visit them and read it aloud. In experiments, the people who made the visits remained happier and less depressed for a month afterward (and the people who heard the letter must have felt better, too).

If all that seems too much work, you could try a once-a-year exercise in gratitude that takes two minutes and costs nothing more than the price of a tablecloth. At Thanksgiving dinner, hand out pens and ask everyone at the table to write on the tablecloth something for which they're grateful. We have no data on the effectiveness of this strategy, but we have anecdotal evidence (from our own Thanksgivings) that it makes for a happier dinner. The effects get stronger every Thanksgiving that you reuse the tablecloth, and it makes for good reading the rest of the year, too.

Make time for nostalgia—and make more good memories. Besides enjoying his memories, Sedikides promotes "anticipatory nostalgia" by deliberately creating moments today that will make for pleasant thoughts in the future. The benefits of events like birthday dinners and weekend trips endure long after they're over. When companies are considering whether to pay for a holiday party, Sedikides suggests they consider a finding from his studies: The more nostalgic memories that employees have of company events, the less likely they are to look for other jobs. In his own life, when Sedikides gets together with friends and family and colleagues, he follows a motto: "Make it memorable."

Treasure the past, but don't compare. Nostalgia isn't useful when it compares the past with the present and invokes a rueful feeling of "Those were the days." It's the pitfall described in "Suite: Judy Blue Eyes" when Stephen Stills sings, "Don't let the past remind us of what we are not now." Psychologists call it "self-discontinuity," a sense of loss and dislocation that has been linked to physical and mental ills. It's better to use the past to make sense of what your life has meant, to see the memories as assets rather than as reminders of what's missing.

When Pollyanna lies paralyzed in bed and says she's glad she once had legs that could walk, it seems unbelievably cheerful—Pollyannaish in the worst sense of the word—but her basic strategy is sound. It's the same one employed by a far more credible character at the end of a far better movie, *Casablanca.* When he says good-bye to Ingrid Bergman on the tarmac, Humphrey Bogart dwells not on what he's lost but on what remains and cannot be taken away. "We'll always have Paris," he says—and he's right.

CHAPTER 9

The Crisis Crisis

Bad Ascending

Whether you're absorbing today's bad news or contemplating the future of humanity, we suggest starting off with three assumptions:

1. The world will always seem to be in crisis.
2. The crisis is never as bad as it sounds.
3. The solution could easily make things worse.

Obviously, there are some real problems that need to be solved. But we're convinced that the biggest problem of all, the greatest obstacle to freedom and prosperity, is the exploitation of people's negativity bias by crisismongers. It's the worst social consequence of the negativity effect. By continually fomenting fears, the prophets of doom have profoundly distorted the public's view of the present and the future. By hyping small or nonexistent threats to induce panicky responses, they create far more problems than they solve.

The result is what we call the Crisis Crisis, although we're not trying to scare anyone into thinking it's a brand-new threat to humanity's survival. The modern barrage of negativity is especially intense and fast paced thanks to the round-the-clock alarmists on our screens, but people have long been susceptible to crisismongers. In 1918, long before cable news and the web, H. L. Mencken described public discourse as "a combat of crazes" and shrewdly diagnosed the fundamental problem in politics and public policy.

"The whole aim of practical politics," Mencken wrote, "is to keep the populace alarmed (and hence clamorous to be led to safety) by menacing it with an endless series of hobgoblins, most of them imaginary."

The specific hobgoblins vary according to ideology, but politicians across the spectrum exploit the same basic cognitive biases. Whether they're spreading alarms about terrorists, immigrants, new technologies, or environmental catastrophes, they agree that things are going to hell because of an unprecedented threat. They yearn to restore the glories of yesteryear, when the citizenry was virtuous, the nation was great, and Earth was pristine. Their gloomy vision of a world in decline makes intuitive sense because we overreact to current negative events (the usual power of bad) while applying a different filter toward history. As mentioned in chapter 8, we tend to see our own past through rose-colored glasses, and we have a similar positivity bias when looking backward at the rest of the world.

We're fooled by what the psychologist Carey Morewedge calls the record-store effect: Our memory is like a record store that stocks a wide variety of current songs but only the greatest hits of the past. His experiments show that people tend to evaluate today's entertainment by comparing it with the best from their youth. If you're assessing the state of music by comparing a random song on the radio today against Nirvana or Pearl Jam, you'll find yourself lamenting the decline from a golden age of music in the 1990s. But the nineties didn't feel so golden at the time. People were listening not just to Nirvana but also to "Achy Breaky Heart" and "I'm Too Sexy"—and comparing those songs with the golden age of the Beatles and the Rolling Stones. Similarly, when we compare today's leaders with Lincoln and Churchill, or today's political squabbling with the grand achievements in history textbooks, we're bound to feel that the nation's best days are past.

This Golden Age fallacy has been deluding people for thousands of years. It's the bad form of nostalgia that we warned against in the previous chapter: looking at the past as a better time that can never be regained. The fallacy was formally introduced in the eighth century B.C. by Hesiod, a Greek poet and farmer who has been called history's first economist. Like today's doomsayers, he bemoaned the dangers of new technology. He explained that humans had once "dwelt in ease and peace upon their lands with many good things, rich in flocks and loved by the blessed gods." But after Prometheus stole fire from Zeus, the gods punished this technological innovation by inflicting humanity with Pandora's box of troubles. The Golden Age gave way to inferior ages of silver and bronze, and ultimately to the current Iron Age, when men "never rest from labour and sorrow." Soon, things would be even worse: "Zeus will destroy this race of mortal men."

Hesiod's obituary for Greece was a bit premature, considering that he was writing centuries before the Parthenon and the Age of Pericles. But he deserves credit for being more realistic than modern doomsayers. The Greeks of his day really did have plenty of woes in their future: conquests by invaders, sporadic plagues and famines, perpetual poverty or slavery for most of the population. That was the fate of people everywhere until industrialization. Humanity was constantly menaced by the Four Horsemen of the Apocalypse: Death, Famine, War, Pestilence. Half of the children died before the age of five. A minor injury or infection could be fatal. A crop failure could mean starvation.

Virtually everyone was poor. The only ways to acquire wealth were to seize it from others or force them to work for you. Serfs were bound for life to feed their lords. Slavery was an accepted tradition on every inhabited continent. Technological progress was slow and sporadic, subject to the whims of rulers who quashed inventions and heresies that threatened the status quo. The emperors of ancient Rome, like those in Persia and India and China, patronized an educated elite but didn't share their wealth and knowledge with commoners. Technological advances in China, like the first mechanical clock and vastly improved methods of smelting iron, were suppressed by mandarins fearful of change.

The standard of living for the masses remained essentially unchanged until a revolutionary set of ideas and institutions arose in Europe. The fall of the Roman Empire left the continent decentralized with independent fiefdoms where scholars, inventors, and merchants could share knowledge and innovate without imperial interference. The medieval era, formerly mislabeled the "Dark Ages" by nostalgists for the leisure-class splendors of Rome, was actually "one of the great innovative eras of mankind," in the words of Jean Gimpel, the French historian. He calls it the first industrial revolution. While the Roman economy had been powered by slave labor, medieval engineers tapped natural forces by building dams and new efficient waterwheels throughout Europe. Windmills proliferated and were used to drain the coastal regions of the Low Countries. Germanic "barbarians" developed a much-improved form of steel. The Vikings made major advances in shipbuilding and navigation. Mechanical clocks and eyeglasses were introduced. Agricultural productivity surged thanks to advances in crop rotation and the invention of the harrow and a new heavy plow, so the average person was better fed and healthier than during Roman times.

Medieval monasteries promoted research and entrepreneurship as they produced and marketed products much as modern corporations do, with the abbots serving as CEOs. Traders and bankers in the city-states of northern Italy led a commercial revolution, an international exchange of goods and ideas that sparked the Renaissance. When the Italian city-states fell under the sway of foreign monarchs, the merchants and artists moved to the Low Countries. Then, after the burghers were stymied by Hapsburg rulers, the entrepreneurial capital shifted to Britain, where the monarch's power was constrained by law, and it was there that scientists, engineers, and capitalists collaborated to start the Industrial Revolution.

This was the first truly golden age: the Great Enrichment, as the economist Deirdre McCloskey has dubbed the astonishing burst of prosperity and expansion of freedom in the nineteenth and twentieth centuries. The average human's income, after stagnating for millennia, increased tenfold in just two centuries. The Industrial Revolution made possible the vast creation of wealth without conquest or enslavement. Philosophers and theologians had long recognized the moral evil of slavery, but the movement to abolish it gained strength only when there were machines available to do the work instead. New technologies were commercialized for the masses, easing burdens for everyone and turning commoners into a bourgeois class with the power to demand new universal rights and freedoms. Two centuries later, the majority of humanity lives in democratic countries that keep growing more prosperous.

The Four Horsemen have never posed so little threat. Death has been forestalled by the greatest miracle in the history of the human species: the doubling of life expectancy since the Industrial Revolution. War still ravages some countries, but as Steven Pinker has documented, we are living in what is probably the most peaceful era in history. Never before has the average person faced such a small threat of dying in war or from other forms of violence. The incentive for future war has diminished because there's no longer an imperative to seize wealth or farmland from neighboring countries. Despite population growth, farmland in North America and Europe has long been reverting to forests and grasslands because farmers can feed more people on less land, and that trend is projected for the rest of the world, too.

Famine and Pestilence afflict fewer people than ever before. Last century it was estimated that only half of the world's people were adequately nourished; today nearly 90 percent are. (The biggest nutritional problem in many places is now obesity.) There has been so much global progress against disease since 1950 that life expectancy in poor countries has increased by about thirty years—the most rapid increase in history. Rates of literacy and education have risen around the world, and people enjoy unprecedented leisure. In the middle of the nineteenth century, the typical man in Britain worked more than sixty hours a week, with no annual vacation, from age ten until he died in his fifties. Today's workers enjoy three times as much leisure over the course of their lives: a gift of two hundred thousand extra hours of free time.

Just about every measure of human well-being has improved except for one: hope. The healthier and wealthier we become, the gloomier our worldview. In international surveys, it's the richest people who sound the most pessimistic—and also the most clueless. Most respondents in developing countries like Nigeria and Indonesia know that living conditions have improved around the world, and they expect further improvement in the coming decades. But most respondents in affluent countries don't share that optimism because they don't realize how much progress is being made.

In the past two decades, the rate of child mortality in developing countries has been halved, and the global poverty rate has been reduced by two-thirds, but most North Americans and Europeans think these rates have remained steady or gotten worse. When they're asked if the world is getting better or worse, or staying the same, fewer than 10 percent say that it's getting better. By an overwhelming majority, they're convinced it's getting worse.

We've escaped the Four Horsemen, but our brains are still governed by the negativity bias. We react in accordance with an old proverb: *No food, one problem. Much food, many problems.* We discover First World problems and fret about remote risks. We start seeing problems even when they aren't there, a propensity that was nicely demonstrated in 2018 by the social psychologist Daniel Gilbert and colleagues.

The researchers showed people a series of colored dots and asked them to decide whether each one was blue or not. As the series proceeded, the prevalence of blue dots declined, but people were determined to keep seeing the same number anyway, so they'd start mistakenly classifying purple dots as blue. The same thing happened when people were shown a series of faces and asked to identify the ones with threatening expressions. As the prevalence of threatening faces declined, people would compensate by misclassifying neutral faces as hostile. In a final test, people were asked to evaluate a series of proposals for scientific studies—some clearly ethical, some ambiguous, some clearly unethical—and reject the unethical ones. Yet again, as the prevalence of unethical proposals declined, people compensated by rejecting more of the ambiguous proposals.

These mistakes happened even when people were explicitly warned ahead of time that the prevalence of the targets would decline, and even when they were specifically instructed to "be consistent" and offered cash bonuses for being accurate. Once they started looking for anomalies, they kept seeing them even as the anomalies were vanishing. Since our brains are primed to spot bad anomalies more readily than good anomalies, we keep imagining new problems even when life improves. As Gilbert notes, "When the world gets better, we become harsher critics of it, and this can cause us to mistakenly conclude that it hasn't actually gotten better at all. Progress, it seems, tends to mask itself."

This "prevalence-induced concept change," as Gilbert calls it, helps explain a trend we've noticed in public discourse. As life expectancy increases and the traditional threats to each person's life diminish, we spend more time worrying about hypothetical dangers to humanity. Our ancestors had good reasons to fear early death, but they generally didn't worry about everyone else perishing, too. We take our personal longevity for granted but imagine that terrorism or nuclear weapons or climate change poses an "existential threat" to civilization or life on Earth. There seems to be an inverse scare law: the more remote the danger, the more apocalyptic the warnings. Our security and prosperity give us more time to worry, more things to fear—and more fearmongers to exploit those feelings.

The Merchants of Bad

The modern world will always be in crisis because its wealth and freedom have created a crisis industry. In the agricultural era, society could afford to support just a few intellectuals, usually beholden to royal patrons who didn't welcome criticism of their policies, and certainly weren't about to pay anyone to alarm the public. The chief prophets of doom were priests with the political savvy to stick to spiritual crises. They warned of God's Day of Judgment or the personal day of judgment awaiting sinners (the hell crisis). But after the Industrial Revolution, as democracy and education spread in Europe, a new class of secular doomsayers emerged armed with charts, theories, and printing presses.

They've grown into an industry of journalists, politicians, and an ever-widening array of "experts" in academia, think thanks, corporations, and nonprofit groups. When something goes wrong, or at least seems to go wrong, they proclaim it an unprecedented threat. They have no patience for analyzing long-term trends, and no faith that people may already have the tools to solve this problem. Catastrophe cannot be averted without the experts' guidance.

They use current events to stoke ancient emotions, like the xenophobia essential to our ancestors' survival. Cities from Ur and Jericho through the Renaissance had a good reason for surrounding themselves with walls: to keep out the foreign armies that were a continual threat. That threat no longer exists in North America, but xenophobic politicians can still whip up enthusiasm for a wall at the American southern border. The United States has been the world's military superpower for seventy years, a remarkably peaceful epoch, but experts have been continually alarmed about a "missile gap" or "window of vulnerability" or other supposed threat to the nation's existence. There was a brief hiatus from crisis in the early 1990s, when the Soviet Union disbanded and agreed to destroy three-quarters of its nuclear arsenal. But even without a credible military rival, America's experts deftly found new cause for alarm. Small countries like Iran and Iraq suddenly became large menaces, and terrorism was promoted to a permanent crisis after the attacks on September 11, 2001. Terrorism's threat was nothing compared with that from the Soviet Union, but spending on national defense and security increased anyway.

When they're not fretting about foreign foes, crisismongers can always find enemies within, often by singling out members of an unpopular group. Our negativity bias has unfortunate implications for one of the most cherished ideas in American social science: the "contact hypothesis," which is the notion that prejudice between groups can be overcome simply by bringing the groups into contact with one another. Social scientists quite reasonably assumed that because prejudice against outside groups resulted from ignorance, it could be overcome once people got to know the outsiders. But when researchers tested the hypothesis by studying communities with high ethnic diversity, they found just the opposite. There were higher levels of prejudice in neighborhoods where different groups lived side by side.

An explanation emerged in recent studies in America and Australia. As predicted in the contact hypothesis, the whites who had the most positive interactions with minorities—blacks, Muslims, and Aboriginal Australians—did indeed become less prejudiced. But while most of the contacts between whites and minorities were positive, people's judgment was skewed by the power of bad. Each individual negative experience had such greater impact that the overall effect was negative: The more contacts you had with minorities, the more likely you were to be prejudiced. If, say, you were a native-born American or Australian living next door to a store run by a family of immigrants, you'd probably have many positive contacts when you shopped at the store. But if you were ever shortchanged, or if the family's teenagers ever caused a disturbance on the sidewalk, that single negative contact would loom large with you—and it would be just the sort of negative gossip that you might share with others in the neighborhood.

Those findings are certainly not an argument for segregated neighborhoods. Other studies have shown that prejudice can be overcome when there are enough pleasant interactions to outweigh the negative ones. The lesson is that people in any neighborhood need to remember and compensate for the power of bad. The same lesson applies when making judgments about the members of any outside group, especially one that's stigmatized by the majority. The negativity bias makes us prone to quickly generalize any individual's offense into animus against the whole group—and then go looking for more offenders. The witch-hunting instinct is easy to arouse.

Historians and sociologists have often explained moral panics as aberrations, attributing them to social upheavals and economic downturns, but in fact they occur in good times as well as bad. In his history of popular delusions, *The United States of Paranoia,* Jesse Walker documents a relentless series of moral panics and wild conspiracy theories, and they're not just perpetuated by fringe groups. They're regularly embraced by the majority and directed at minorities. Mainstream clergymen and journalists told lurid stories about demonic Methodists and Shakers in the eighteenth century, and equally demonic Catholics and Mormons in the nineteenth century. Today, of course, the villains are Muslims. The latest wave of immigrants has always been a threat to the American way, whether they're German, Irish, Italian, Chinese, Japanese, Latino, or Middle Eastern.

Once a moral panic begins, politicians follow that classic bit of illogic: *Something must be done. This is something. Therefore it must be done.* Remarkably little attention gets paid to whether the solution will do any good—or whether history suggests it will make things even worse. Consider the recent reaction to the increase in opioid-related deaths. There have always been people prone to substance abuse, but instead of focusing on treatment for them or addressing the underlying social causes, politicians and journalists keep trying to solve the problem by eliminating whatever substance happens to be in vogue. The past campaigns against marijuana, heroin, cocaine, methamphetamines, and other drugs have sent millions of Americans to prison and cost more than $1 trillion, but the drugs have become easier to obtain than ever—the street prices have been declining for decades. Meanwhile, the drugs have become more dangerous, because the logistics of the black market encourage traffickers to sell more compact and potent varieties (like the bathtub gin that replaced beer during Prohibition). As a result, the economists Christopher Coyne and Abigail Hall conclude, America's long war on drugs has led to an increase in drug overdoses.

But those lessons are being ignored during the panic over opioid-related deaths. The negativity effect rules. Prescription opioids provide enormous benefits to patients—and there are at least fifty million Americans suffering from chronic pain. But journalists (with a few exceptions, like Jacob Sullum in *Reason* and Maia Szalavitz in *Scientific American*) prefer to obsess on the small fraction with a substance-abuse disorder. Politicians of both parties have voted to restrict prescriptions. Their perception of opioid use was summed up by President Donald Trump: "It's so highly addictive. People go into the hospital with a broken arm; they come out, they're a drug addict."

In reality, the risk of a patient becoming addicted (much less dying) is probably less than 1 or 2 percent, according to medical-journal studies tracking patients and to the federal government's annual National Survey on Drug Use and Health. The survey has repeatedly shown that most people who misuse prescription painkillers don't get the pills from their doctor. The typical victim of a fatal overdose is someone with a history of alcohol or drug abuse, often both, and often a history of mental illness, who obtained the painkiller illegally and died by combining it with alcohol, a tranquilizer, cocaine, heroin, or another dangerous substance. The New York City Department of Health's statistics show that 97 percent of fatal drug overdoses involve at least two substances.

The crackdown on opioid prescriptions has made it much harder for patients to get proper treatment for their pain, but it hasn't stopped others from abusing opioids. With fewer prescription pills available on the black market, they've switched to much deadlier substitutes like heroin and fentanyl. So as the total volume of opioid prescriptions fell by nearly a third between 2010 and 2017, the number of opioid-related deaths more than doubled. The toll from crisismongering amounts to twenty-five thousand additional deaths annually—and millions of patients in pain suffering needlessly.

Another reliable source of business for the crisis industry is technophobia, which is why fearmongers are always seeing the next Pandora's box. In the nineteenth century, the *New York Times* fretted that electric lights would injure people's retinas, and British doctors warned that riding a train caused brain damage (from the jolts and vibrations, which were supposedly so harmful that a passenger could suddenly turn into a violent "railway madman"). Today's alarmists, with no more convincing evidence, warn that mobile phones and genetically modified foods will cause cancer. They invoke the precautionary principle, a concept that rules federal agencies like the Food and Drug Administration and has been enshrined in laws of the European Union and the United Nations: No technology should be introduced unless it can first be proven to be safe. The formal principle wasn't articulated until the 1980s, but it's essentially the same philosophy followed by the Chinese mandarins who suppressed mechanical clocks, and it's just as antithetical to progress. As the science journalist Ronald Bailey has observed, the most accurate way to state the precautionary principle would be: *Never do anything for the first time.*

Alarmists appeal to old tribal instincts by pitting ruralites against urbanites—the current Red-Blue divide in America—and by promoting class warfare, another development made possible by the Industrial Revolution. It was tough to exploit class divisions during the long period when everyone was poor. But as the Great Enrichment lifted incomes and created new social strata, there were new fears and resentments to tap. The writer William Tucker observed that there have always been two successful forms of demagoguery: telling poor people that rich people have too much money, and telling rich people that there are too many poor people. The first was famously exploited by Marx, who identified the "crisis of capitalism" and inspired its elimination in communist countries (where the Great Enrichment promptly ceased, leading to famines that killed more than fifty million people). The second has been exploited by a series of intellectuals starting with Thomas Malthus, the British clergyman concerned by the population boom in England as life expectancy rose during the

Industrial Revolution. In 1798 he announced with mathematical certainty that humanity was doomed to mass starvation because population growth would inevitably exceed the growth in food production.

The opposite has occurred, but that hasn't discouraged other scholars from making the same mistake over and over. Their negativity bias blinds them to good long-term trends. As the Great Enrichment led to a twentieth-century boom in population in less developed countries, there were predictions of global famines in bestselling books starting in the 1940s with *Our Plundered Planet* and *Road to Survival.* In 1972 a group of prominent economists and scientists called the Club of Rome produced computer models projecting a global population crash due to shortages of food and energy. That scenario was popularized by the ecologist Paul Ehrlich, the bestselling author of *The Population Bomb* and *The End of Affluence,* who in 1970 feared that four billion people, including sixty-five million Americans, would perish in the "great die-off" of the 1980s.

Ehrlich and a fellow academic, John Holdren, wrote in their 1977 ecology textbook that government had the right to limit family size in order to avert mass starvation. The book discussed methods of "involuntary fertility control," including the implantation at puberty of a "sterilizing capsule" under a woman's skin that could be removed only with "official permission." As an intellectual enterprise, the "population crisis" was largely a Western phenomenon, promoted by academics, journalists, activists, and donors in the United States and Europe, but the consequences were suffered elsewhere. The reckless doomsaying inspired some of the worst human rights violations of the twentieth century. Millions of women in Latin America, Bangladesh, and India were sterilized, often coercively. China forbade women from having more than one child, a policy that resulted in tens of millions of coerced abortions.

When the global famine didn't arrive on schedule, Ehrlich and Holdren pushed back the date. In the 1980s they warned that famine caused by climate change could kill a billion people worldwide by 2020, and Holdren kept insisting this remained a possibility when asked about it during his 2009 confirmation hearing to be the White House science adviser. Today Holdren's foresight is looking as bad as ever. Since 2010, the annual number of famine victims worldwide is estimated to be less than forty thousand—the lowest number in recorded history even though the global population has grown sevenfold since Malthus's day. (And the recent famine deaths, in Africa, were due not to a widespread shortage of food but rather to civil wars in which enemies deliberately starved each other.)

Holdren's mistaken predictions and readiness to eliminate a fundamental human right didn't prevent him from becoming the presidential science adviser. They didn't even attract much attention. Journalists extended him the usual professional courtesy to a fellow doomsayer. Errors are rarely career killers in the crisis industry, as the political scientist Philip Tetlock found by systematically tracking nearly thirty thousand predictions by academics, columnists, and others paid for their public prognostications. He reported that these professionals' accuracy rate was barely better than chance (a "dart-throwing chimp," as he put it) and actually worse than that of some nonexperts. What experts lack in accuracy they make up in confidence and exaggeration, which is why they keep getting quoted. The crisis industry is a codependency of journalists, politicians, and the experts feeding them scary predictions. The journalists need sensational stories; the politicians need campaign issues; the experts need publicity, prestige, and funding.

We've called these groups the merchants of bad, but we don't mean they're all in it just for the money. Many of them are genuinely alarmed. The most effective doomsayers are the ones who believe their own prophecies. Chicken Little was truly convinced that the sky was falling. The issue wasn't her sincerity but her misinterpretation of the acorn that fell on her head and her plan for dealing with it.

She and the other animals were so fearful of this imaginary hobgoblin, in Mencken's words, that their clamor for safety led them to seek shelter in the den of the Fox, who promptly made a meal of them. That's the cautionary lesson from the story, and it applies to the Crisis Crisis. There are a lot of hungry foxes out there.

Collective Stupidity

Even if the prophets of doom are exaggerating, don't they still serve some purpose? After all, the negativity effect makes losses loom large, so shouldn't we take extra precautions even if they turn out to be unnecessary? What if the prophets are right this time? That's the crisis industry's standard rationale. But there's a larger problem to consider: a "social stupidity," in the words of Mancur Olson.

Olson was a brilliant American economist who identified the greatest obstacle to prosperity in democratic societies. He studied collective-action problems, in which individuals pursuing their own rational self-interest end up harming the entire society—a social stupidity. His most famous book, *The Rise and Decline of Nations* (1982), analyzed the puzzling pattern of economic growth after World War II. Why did the losers, Germany and Japan, recover so quickly and unexpectedly that it was called an "economic miracle"? Why did one of the winners, Great Britain, become known as the "sick man of Europe"? Once the powerhouse of the Industrial Revolution, it grew so slowly in the twentieth century that economists looked for causes of "the British disease." Olson diagnosed it as a hardening of the economic arteries—"demosclerosis," as it was later dubbed by the author Jonathan Rauch—caused by the gradual accumulation of favors to special-interest groups.

Olson showed how countries suffer when groups of industrialists, unions, farmers, professional societies, and other special interests use their power to fix prices, inflate profits and wages, and lobby for subsidies, tariffs, tax loopholes, and regulations to restrict competition. The results are terribly costly to society, particularly to the poor, who don't have powerful special-interest groups to defend them. But a social stupidity is at work: The average citizen has no personal motivation to battle any single lobby. Unless there's a major disruption, such as the war that destroyed Germany's and Japan's power structures and gave their economies a fresh start, the special-interest groups go on gradually enriching themselves at the public's expense.

"It's death by a thousand cuts," Olson explained. "A kind of status-quo bias gets built into society by all these powerful groups. The rules discriminate against new people, new companies, new ideas—precisely where the dynamic forces of a society should be coming from. So a kind of institutional sclerosis sets in, a clogging of the national arteries. I think it's the single most important problem in societies. It's hard to see, because no single one of these social stupidities is fatal, but enough of them can eventually do a society in."

Crisismongering is an integral part of demosclerosis, Olson recognized, because it's exploited so readily by politicians and special-interest groups. They follow the dictum famously articulated by Rahm Emanuel, the former Chicago mayor and White House chief of staff: "You never want a serious crisis to go to waste." Society as a whole doesn't benefit from overreacting to hyped or nonexistent problems, but it's not in the self-interest of an ordinary citizen to debunk a crisis. So Chicken Little faces little skepticism, and there's usually at least one special-interest group ready to play the role of the Fox by suggesting a quick solution.

The crisis may not be real, and the solution may not do any long-term good, but politicians are in too much of a hurry to find out. They can reap immediate publicity and other benefits (like campaign contributions) by rushing to the rescue. Whatever policies they enact, the effects will linger long after the crisis has passed. When a nation expands its army to fight a war, the military establishment will not shrink back to its former size afterward. As the economist Robert Higgs chronicled in his history of government growth, *Crisis and Leviathan,* bureaucracies and regulations and subsidies do not disappear when the a crisis is over; nor do the special interests that benefit from those policies.

The costs of crises add up, and they aren't just monetary. We spend so much time and energy worrying about small or imaginary threats that we end up being less safe. Like the ancient emperors who outlawed new technologies, modern fearmongering slows the sort of innovation that produced the Great Enrichment. Agricultural scientists, whose Green Revolution in the mid-twentieth century fed the population boom in Asia, expected a similar revolution in African agriculture through the use of genetically modified crops, but this second green revolution has been stymied by Greenpeace and other anti-GMO activists in Europe and America. Besides scaring the public (so that half of Americans think GMOs are unhealthy), their campaign has led foundations and governments to stint on funding for GMO research, and it has prevented farmers in Africa from using crops that would feed more people and combat nutritional deficiencies. The fearmongering even caused Zambian officials to reject foreign food aid during a famine because it contained GMOs.

It's estimated that more than one million people die each year, and at least 250,000 children go blind, because of deficiencies in vitamin A and other micronutrients that could be alleviated with fortified varieties of rice, bananas, and other foods already developed with genetic techniques. These crops are considered safe by an overwhelming majority of scientists, including more than 140 Nobel laureates who have called the anti-GMO campaign a "crime against humanity," but the opinion gap between scientists and the public on this issue remains even larger than the gaps on evolution and climate change. The laureates don't have Greenpeace's budget for lobbying and publicity. Nor do the children who are going blind and dying from vitamin A deficiency.

Medical researchers are primed for their own revolution thanks to advances in genetics and new digital technologies, but they, too, find themselves stymied by new obstacles: the enormous regulatory barriers that have gradually built up in classic demosclerotic fashion. Whenever there's a problem with a new drug or technology, the victims are obvious—and easy fodder for journalists and trial lawyers to use in proclaiming a new crisis, which prompts regulators to add safeguards for future drugs. But the biggest medical scandal of all is the cumulative effect of all these safeguards, which is to slow development of treatments for obesity, diabetes, Alzheimer's, and other diseases. In the past quarter century, the average inflation-adjusted cost of developing a single new drug has increased sixfold. The overall price tag is now $3 billion (some estimates put it at $6 billion), and the approval process typically takes more than a decade and one hundred thousand pages of paperwork. This scandal occasionally attracts attention if there's a well-organized group affected, like the AIDS patients who succeeded in cutting some of the red tape that was delaying the introduction of drugs that could save their lives. But the problem is usually ignored because there aren't readily identifiable victims.

How many people have died waiting for lifesaving drugs to be developed or approved? No one knows the size of this "invisible graveyard," as medical economists call it, but there's no doubt it's huge. You can get a sense of the scale from one example of how deadly fearmongering can be: the hundreds of thousands of Europeans and Americans who die annually because of regulations and technophobia involving nicotine.

Nicotine has a bad reputation because it's linked to cigarette smoking, which is rightly feared as the leading preventable cause of death. But nicotine isn't what makes cigarettes fatal. Smokers are killed by the thousands of toxins and carcinogens in cigarette smoke. Nicotine in itself is "no more harmful to health than caffeine," as Britain's Royal Society for Public Health puts it. Like caffeine, it creates dependence but provides significant benefits. Nicotine has been shown to help people control their weight, to reduce anxiety, and to improve concentration, mood, memory, and alertness. That's why so many people go on smoking even though they know how dangerous it is. Despite a half century of exhortations to quit, at least 15 percent of adults still smoke in every European country except one.

The sole exception is Sweden, where the smoking rate has fallen to just 7 percent because so many people have turned to an alternative source of nicotine called snus, a form of smokeless tobacco absorbed orally from a small packet. Thanks to this cigarette substitute, Sweden has the lowest rates of smoking-related diseases in Europe, and it's estimated that 350,000 lives would be saved annually if the rest of Europe followed Sweden's example. But instead of encouraging this trend, the European Union has banned snus everywhere except Sweden.

How could public officials deprive people of a lifesaving product? Because of fears fomented by an unlikely alliance of special-interest groups. It's the type of alliance, commonly deployed in the Crisis Crisis, that economists call a Baptist-bootlegger coalition. The term comes from the Baptist preachers who campaigned for blue laws and the bootleggers who wanted liquor stores closed on Sunday so that they could sell alcohol illegally. The Baptists provided politicians with a virtuous excuse for increasing the profits of the bootleggers (and also an opportunity to collect kickbacks from the grateful bootleggers).

In the coalition against nicotine, the role of the virtuous Baptists is played by the hard-liners in the antismoking movement who are opposed to nicotine in any form, insisting that abstinence is the only acceptable strategy. The bootleggers are the industries that benefit by suppressing competition from smokeless tobacco: cigarette companies (for obvious reasons) and pharmaceutical companies that sell nicotine-replacement therapies like nicotine gum, lozenges, and patches—none of which has proved popular with smokers trying to quit.

This coalition, with the usual help from alarmist journalists, has created a popular conception that smokeless tobacco is dangerous even though researchers have concluded that it is at least 99 percent safer than smoking. Long-term studies of snus users have found little or no increased risk of oral cancer, nor any increase in the risk of heart attacks, strokes, or any other illness linked to smoking. But the fearmongering, plus adept lobbying behind the scenes, has kept snus illegal in Europe and prevented snus merchants in the United States from advertising how much safer their product is than cigarettes.

Now this same coalition has turned its attention to suppressing competition from an even bigger threat to their businesses: electronic cigarettes, which deliver nicotine in a vapor without the toxins in tobacco smoke. These vaping sticks have been shown, in a rigorous study in 2019, to be twice as effective as other nicotine-replacement products (like gum or patches) in helping smokers quit. Since their introduction in 2010, smoking rates in America have fallen sharply, dropping for the first time below 15 percent for adults. One popular vaping stick, the Juul, has proved so effective at weaning smokers from their habit that Wall Street analysts have called it an existential threat to the tobacco industry, whose stocks have plunged along with cigarette sales.

This technology has the potential to save tens of millions of lives worldwide, yet regulators around the world, in response to lobbying by the Baptist-bootlegger coalition, have taken steps to outlaw or severely restrict vaping sticks. They've justified this crackdown by invoking the precautionary principle: Even though there haven't been any demonstrated harms from nicotine vapor, we don't yet know for sure that this new technology is absolutely safe. That's true, but it just shows how deadly the precautionary principle can be. Whatever harms might eventually emerge from vaping, they're of "minimal consequence" compared with those from smoking, according to Britain's most eminent medical authority, the Royal College of Physicians, which concluded that vaping is at least 95 percent safer than smoking. Worrying about tiny hypothetical dangers tomorrow instead of a large real danger today is an absurd form of safety addiction—negativity bias at its most neurotic.

Yet that's been the approach not only of regulators but also of the press, which has kept running stories about hypothetical harms from e-cigarettes. Journalists have fixated on an "epidemic" of vaping among American teenagers and claimed it acts as a "gateway" to smoking, but meanwhile the rates of youth smoking have plummeted to historic lows, declining much more steeply than during the prevaping years. The majority of teenagers who try the sticks don't even become regular vapers, much less smokers. (Many are using liquids without any nicotine.)

No one wants to see teenagers acquire a nicotine habit—the vaping industry supports bans on sales to minors—but that risk doesn't justify the restrictions (like the bans on flavored vaping liquid) that are discouraging smokers of all ages from quitting. There are a lot more teenage drinkers than vapers, but we don't restrict sales of alcohol to adults even though it poses far more dangers to teenagers. Vaping is not only safer but also provides enormous benefits: Since the emergence of vaping, the smoking rate among teenagers and young adults has been cut in half.

You'd think this development would be welcomed by antismoking groups—and it has been by some—but most of them have developed a virulent case of what's called the March of Dimes syndrome. It's a common condition in crisismongering groups, although it's named after a much worthier institution. The March of Dimes, created to fight polio, did not declare victory and disband once the polio vaccine largely ended that threat. It found new diseases to fight. Antismoking groups and public-health officials are in a similar situation. They've largely succeeded in their mission to educate the public about the hazards of smoking and make it illegal in most public spaces. What's left for them to do now that smoking is universally reviled? How can they attract media attention to a threat that's such old news?

They need a new threat to justify their jobs, and unfortunately the most convenient one is vaping. Their scare campaign has been good for the budgets of public-health officials and activists, but it has been a disaster for the public's health. Surveys over the past decade show a reversal in popular opinion toward vaping. It was initially considered safer than smoking, but now the prevailing view among Americans is that it's just as harmful—or even more harmful. Europeans are similarly misinformed. That means that millions of smokers have been dissuaded from making the switch, and many of them will end up in that invisible graveyard of victims killed by the Crisis Crisis.

We could go on toting up the costs of social stupidities, but we don't want to end up sounding like doomsayers ourselves. We don't want to focus so much on individual problems that we miss the larger picture. Despite all the unnecessary obstacles to progress, humanity continues to get healthier, wealthier, and wiser. We really are living in a golden age, even if most people believe otherwise. The Great Enrichment continues, and it will proceed even faster if we can overcome the one real crisis by learning to ignore the fearmongers.

We need to close the gap in public-opinion surveys showing that most people are pessimistic about the world but optimistic about the prospects for themselves and their own community. Researchers call it the Optimism Gap, also known as the "I'm OK, They're Not" syndrome. In America and other affluent countries surveyed, most people expect their own financial situation to improve in the next year, but they're pessimistic about the economy of their nation and the world. When asked about topics like immigration, unemployment, teenage pregnancy, crime, and illegal drugs, most people say these aren't a problem in their own community but are a serious problem for their country. They're much more satisfied with environmental quality in their own area than with the quality nationally, and still less satisfied with the quality globally.

Not surprisingly, researchers have found that the Optimism Gap is related to the crisis industry: The more time and attention you devote to television news, the more pessimistic you tend to be about the world. How could you not be? No matter how good your day, roving correspondents will find—or fabricate—some fresh horror to disillusion you. No matter how well you use the strategies in this book to cope with the negativity bias in your personal life and career, you still have to contend with the Crisis Crisis. That's a different sort of challenge, and it requires different strategies.

Cutting the Profits of Doom

Why is Washington, D.C., surrounded by the richest counties in the United States? Because demosclerosis has clogged the capital with special-interest groups. The federal government has been expanding for decades, through Republican as well as Democratic administrations, enriching the legions of lobbyists, lawyers, contractors, regulators, journalists, and assorted experts who have gathered there. This concentration of wealth and power is regularly denounced in the rest of the country, to little effect. Because the Crisis Crisis is a collective-action problem, the typical individual has no incentive to debunk the doomsayers or resist the growth of power in Washington, while journalists and lobbyists and the rest of the crisis industry have every incentive to keep stoking fears. They can't be expected to give up their jobs voluntarily.

But we could at least try altering the incentives for doomsaying. We need more rewards for politicians when they speak rationally about risk, as President Barack Obama did when he declared that terrorism is not an existential threat. We need to encourage analysts who put problems in perspective, as the Copenhagen Consensus does by enlisting prominent economists to rank the most cost-effective ways to help the world's poor. (In their list of priorities, slowing global warming ranks much lower than fighting disease and nutritional deficiencies.) And, since penalties are more effective than rewards, we need ways to cut the profits of doom. Here are a few suggestions:

CrisisCrisis.com. As much as journalists love horror stories, they've been forced to cut back on one particular genre: the urban legend. They used to casually allude to classics like the alligators lurking in New York sewers, the rodent served in fast-food chicken (the Kentucky Fried Rat legend), and the Halloween trick-or-treater given poisoned candy or an apple with a hidden razor blade. Those particular legends were all false, but who knew that? As reporters like to say, not entirely facetiously, some stories are too good to check.

But now it's easy to look up those tales at Snopes.com, a website that systematically investigates urban legends. It keeps a repository of the popular ones, labeling them true or false, that's readily available to journalists—and to their readers, who can instantly embarrass them online by posting a link to the Snopes evidence debunking the legend. The smackdown is effective because Snopes has built a reputation for accurate research. It has also built an audience, becoming one of the more popular sites on the web, which shows there can be a market for debunking.

We'd like to see a similar repository evaluating prophecies of doom. Crises aren't as simple to debunk as urban legends—there isn't one story to check—but a little context can do a lot to quell alarmism. Terrorism hardly seems like an existential threat once you see a chart showing that there were far more attacks and fatalities in the 1970s and 1980s (the heyday of the Irish Republican Army and Basque separatists in Spain) than today. The prospect of polar bears going extinct (a prediction ritually accompanied by a photo of an emaciated bear on a melting ice floe) seems implausible once you look at population trends (their numbers have increased in recent decades) and the fact that they previously survived in a climate hotter than the most dire global-warming forecasts. This sort of information is now scattered around the web at sites that promise in-depth analysis of issues, but they're mostly aligned with a particular ideological camp, so their analyses are dismissed by the other side. What's needed is a repository like Snopes.com, one that has a reputation for neutrality and gathers all the information in one easy-to-use site. This repository would be a public service, and it could put the power of bad to good use by including a scorecard for prophets and prophecies. Just as the negativity bias inclines us to pay attention to scary predictions, it also inclines us to pay extra attention to mistakes. Doomsayers would lose some of their power if people could see how often they've been wrong.

Enact Patty's Law. After her eleven-year-old son, Jacob, was raped and murdered near their home in Minnesota, Patty Wetterling became a leader in the movement to mandate public registries of sex offenders. A federal law named after Jacob was passed in 1994, and she returned to Washington two years later when its requirements were expanded by an amendment named after another murdered child, Megan Kanka. She was standing near President Clinton in the Rose Garden when he signed the legislation known as Megan's Law. But today she has a new cause: undoing the harm done by these laws. She has become an outspoken critic of the draconian provisions that have ensnared countless people who don't fit anyone's idea of a sex offender, like the nineteen-year-old who had sex with the seventeen-year-old girl whom he went on to marry, or teenagers whose reputations are ruined for life because they sent a text message to another teenager.

Unfortunately, politicians haven't learned from the mistake of Jacob's and Megan's laws. Instead, they've become even more determined to exploit personal tragedies by hastily drawing up legislation named after victims. Donald Trump campaigned for Kate's Law, named after a woman in San Francisco murdered by an immigrant, and there have been dozens of other victims commemorated in federal and state legislation. It's emotional blackmail. Personalizing a law garners publicity (the victim's family can appear at the news conference) and helps speed passage (who would dare vote against a sympathetic victim?), but it short-circuits rational debate and makes for bad public policy. So in honor of Patty Wetterling, we'll fight fire with fire by proposing one final bit of personalized legislation: Patty's Law, which will forbid any future law from being named after a child or anyone else.

The Red-and-Blue-Ribbon Panel. In the era before round-the-clock news, a common response to a crisis was to appoint a blue-ribbon panel to investigate. By the time its voluminous report came out, the crisis was often such old news that the recommendations were ignored. The blue-ribbon panel became a byword for delay and inaction, and a frequent object of mockery from impatient journalists.

But from the public's standpoint, the delay was a feature, not a bug. It prevented bad ideas from being rushed into law. There's ordinarily a brief window to exploit popular anxiety, but the public's clamor for radical action soon subsides. The blue-ribbon panel sounds rather quaint today, given journalists' incessant need for news and politicians' endless desire for face time on television, but there's no reason the tradition couldn't be revived and improved.

Instead of succumbing to the something-must-be-done impulse by enacting a law, appoint a commission, and don't just stack it with the usual experts from the crisis industry, who are always happy to recommend more funding for themselves. Instead, set up rival teams, employing a format that has been effectively pioneered in the public and private sectors to plot military strategy, design spacecraft, improve cybersecurity, and manage other projects. It's called a red team exercise. A group called the blue team analyzes the problem and develops a solution, which is then critiqued by a red team that's looking for flaws. A panel of referees moderates the back-and-forth debate and eventually produces a well-scrutinized proposal for dealing with the crisis—assuming that by then anyone still cares about the problem.

Put your money where your doom is. When an expert confidently predicts disaster, the first question to him or her should be: Wanna bet? This technique was famously deployed in the journal *Science* by Julian Simon, an economist frustrated by the apocalyptic predictions during the energy crises of the 1970s. Conventional wisdom had it that the price of oil and other natural resources would soar as the growing population's demand outstripped the planet's dwindling supplies. Unlike the doomsayers, Simon had studied long-term trends and knew that the price of energy and other natural resources had been declining for centuries thanks to people's ingenuity at finding new supplies and substitutes. In *Science,* he offered to bet that oil or any other natural resource would be cheaper at any date in the future. Paul Ehrlich, the ecologist, and John Holdren, the future White House science adviser, accepted the challenge. In 1980 they picked five metals (copper, chromium, nickel, tin, and tungsten) and bet $1,000 that their prices would be higher ten years later. When the bet came due in 1990, all five metals were cheaper, and the doomsayers had to pay up, publicly acknowledging their mistake.

Simon's tactic has been adopted by others, including Tierney, who in 2005 placed a $5,000 bet with the author of a cover story in the *New York Times Magazine* predicting, yet again, that oil prices were about to soar as the world's supplies ran out. (Tierney collected his winnings in 2010.) This practice has been institutionalized at Long Bets, a website that calls itself "The Arena for Accountable Predictions." Established by the ecologist Stewart Brand and the writer Kevin Kelly, with financial backing from Jeff Bezos of Amazon, the website allows people to post a prediction on any topic and invite a wager. The biggest one so far was made in 2007 by Warren Buffett, who bet hedge-fund managers that an ordinary index fund of S&P 500 stocks would outperform the hedge funds over the next ten years. Buffett won $2.2 million (and gave it to charity).

Most of the wagers at Long Bets have been for less than $2,000, but it's not the amount of money that matters. The benefit of betting is that it forces people to make specific, testable predictions—and pay a price if they're wrong. If doomsayers want society to spend large sums dealing with a threat, they should be willing to put their own cash—and reputations—on the line.

Don't memorialize terrorism. The impulse to honor the dead is noble, but treating victims of terrorist attacks as a special class of martyrs serves only to create more victims in the future. The candlelight vigils, the national moments of silence, the solemn anniversary ceremonies and stone monuments—they're all well-intentioned, but they inadvertently glorify terrorism, stoking public fears and encouraging future attackers. To a jihadist wannabe, the 9/11 Memorial in downtown Manhattan offers fantasies of his own immortal fame. The victims' families deserve sympathy and solace, of course, but then so do the families of people killed in accidents and other tragedies, and we don't hold anniversary ceremonies or build public monuments for them. The best way to honor a victim of terrorism is to find the killers, punish them, and then forget them.

Stop giving prizes for terror pornography. When a terrorist bomb explodes or a sociopath goes on a rampage, two words promptly pop into the mind of a newspaper editor: Pulitzer Prize. Television news producers have visions of a Peabody Award. They know how regularly the prize judges reward saturation coverage of these tragedies, so they deploy every able body to cover every conceivable angle—precisely the publicity that the terrorists sought. A team of reporters produces a long profile of the sociopath, complete with stories from his childhood, rants from his Facebook page, and photos of him posing as an armed commando—just the sort of glory he craved. The publicity inspires other sociopaths to make their own plans, resulting in a "media contagion" that researchers have identified as a major cause of mass shootings. Researchers have joined with parents of victims and the FBI in a couple of organizations—Don't Name Them and No Notoriety—urging journalists to avoid frequent mentions of the killers' names and refrain from publishing their self-promotional photos and manifestos.

> A few journalists have gone along with these recommendations and acknowledged their role as terror publicists, but most can't resist saturation coverage even when there's little interesting left to say. We realize they're under competitive pressure—no one wants to get scooped on a dramatic story—but we don't see why the Pulitzer and Peabody judges keep rewarding them for fearmongering. The judges are under no pressure to maximize web traffic. They're supposed to be giving prizes to journalism that enlightens and serves the public good. Terror pornography does neither.

Reducing the profits of doom would elevate public discourse, but the merchants of bad will stay in business as long as there are customers. Ultimately, the best way to discourage terror pornography or any other form of crisismongering is for people to stop paying attention. That may sound hopelessly utopian, given the mass appeal of the power of bad and the round-the-clock opportunities for doomsaying. The Digital Revolution has made it easier than ever to spread misinformation and fear, and it has indeed made some people more anxious and angry at one another.

But, as usual, the future isn't as dire as you've heard.

CHAPTER 10

The Future of Good

Bad will always be stronger than good, but good's prospects are improving. For all the harms we've cataloged from the negativity effect, all the mistaken judgments and needless acrimony and social stupidities, we're convinced that bad is more vulnerable than ever because we understand it more clearly than ever. The doubling of life expectancy during the Great Enrichment was made possible by a new understanding of the natural world, which enabled engineers and scientists to harness its forces and overcome its dangers. As we gain understanding of our inner nature, we can work to harness and overcome the negativity effect.

Just as we've staved off the Four Horsemen, we can use our prefrontal cortex to tame the primal impulses of our ancient brain—or at least train ourselves to recognize how those negative impulses warp our perceptions and decisions. The Rule of Four, as we've said, is not a universal law of nature, but it's a useful rough guide. If you remind yourself that it usually takes four good things to overcome one bad thing, you'll know not to trust your immediate responses to bad events. Instead of catastrophizing when there's a setback, you can rationally weigh it against the progress you've made. Instead of assuming the worst when your partner lets you down, you can force yourself to see it from their perspective or rely on someone's else perspective. When you do something wrong, you can understand why a mistake that seems trivial to you makes so much difference to your partner or your family or your boss.

If you remember the Negative Golden Rule—it's what you *don't* do unto others that matters most—you can avoid grief and save energy. Avoiding broken promises will do much more for you than going the extra mile. Instead of striving for perfection, be good enough. By recognizing the impact of criticism—and the ineffectiveness of the criticism sandwich—you can deliver it more usefully, and you can receive it more constructively, too, so that it teaches rather than devastates. By recognizing how much more powerful penalties are than prizes, you can be a better parent or teacher or manager.

Once you appreciate the irrationality of safety addiction, you're liberated to take new risks. It's not easy to do, which is why football coaches keep punting on fourth down, but some of them are getting a little more adventurous, and we expect more will eventually learn from Kevin Kelley's success. Overriding gut responses takes effort, but the techniques used by Fearless Felix to quell his panic are being used by others and are available to anyone. Dealing with problem employees and angry customers takes effort, too, but businesses are learning to avoid bad apples and follow the Casablanca Hotel's strategies for eliminating the negative. The online negativity effect is forcing companies to provide better service to their customers—who then, naturally, find new things to complain about and force the companies to do an even better job of serving the public.

Now that psychologists have recognized the negativity bias that skewed their own research for so long, they've found new strategies for accentuating the positive. They've realized that we're not stuck on a hedonic treadmill limiting our emotional range. By observing how most victims of trauma emerge stronger, and how most people get happier in old age, psychologists have confirmed that versions of Pollyanna's Glad Game really do work. People of all ages can counter the power of bad by consciously rewriting their narratives, focusing on their blessings, and savoring the good moments of their lives. Nostalgia is a tool not just for appreciating the past but also for brightening both the present and the future. Researchers have found that the positive psychology movement can benefit the body as well as the mind, as evidenced by the reduced rates of cardiovascular disease and overall mortality in people who are satisfied with their lives and optimistic about the future.

These happy findings, naturally, haven't attracted as much attention as the bad news. But we're optimistic that can change, too.

The Low-Bad Diet

Every new communications technology is greeted with suspicion, particularly by the political and intellectual elite. Socrates complained that writing weakened people's ability to memorize. The rulers of the Ottoman Empire banned the printing press, which threatened the religious establishment as well as the jobs of traditional scribes, while the leaders of the Catholic Church banned heretical books. Newspapers were blamed for socially isolating people. Radio was denounced as a tool for demagoguery, and television was written off as a "vast wasteland."

When that wasteland began changing in the 1980s and 1990s, giving way to hundreds of cable stations and millions of websites, there were more predictions of disaster from technophobes. The Information Revolution was turning America into a "totalitarian technocracy" with no "transcendent sense of purpose or meaning," according to a much-quoted professor of "media ecology." Books like *Data Smog* and *Amusing Ourselves to Death* warned of "information overload" and bemoaned the crassness of the new media: the shopping channels hawking cubic zirconia rings, the endless infomercials, the shows featuring psychics, astrologers, and porn stars.

The doomsayers inevitably got the most attention, but there was one notable optimist, George Gilder. In 1990, when the Internet was still confined to a few geeks, he published a book titled *Life After Television.* The title sounded preposterous at the time, considering the huge audiences for both network and cable television, but Gilder was gleefully anticipating the downfall of mass media. As the Internet came of age in the mid-1990s, he predicted a golden age of entertainment.

"When I debate network television executives," Gilder said in 1997, "they come up to me afterward and say: 'Look, you don't understand. People like the stuff we put on. We've done market surveys and we've found that people are boobs.' Well, I'm a boob when faced with conventional TV. It's too much work to find something good, and if you do, it probably won't accord with your particular interests. Television is a lowest-common-denominator medium even with 50 or 500 channels. You'd never go to a bookstore with only 500 titles. In the book or the magazine industries, 99.7 percent of the stuff is by definition not for you, and that's what the Internet is like. It's a first-choice medium with a bias toward each individual's area of excellence instead of the few commonly shared interests, which are mostly prurient and sensational."

Gilder's optimism was vindicated. As cable channels proliferated and the web matured, allowing people to watch whatever they wanted whenever they wanted on platforms like Netflix, Amazon, and YouTube, network TV gave way to a golden age of sophisticated dramas, comedies, and documentaries. Sure, there was still plenty of junk, but now there were epic series more rich and complex than anything ever seen on any screen. Television surpassed cinema as the medium for innovation once network executives lost their stranglehold over programming. People didn't just want to watch formulaic sitcoms or buy cubic zirconia rings. Given a choice, they could appreciate excellence.

Now this same sort of evolution is occurring with social media. Freed from the old mass-media news oligopoly, people are finding their own sources of news, and once again there are predictions of disaster. Pundits and politicians decry the rise of "fake news" that's supposedly swinging elections, and the "ideological silos" and "echo chambers" that are supposedly polarizing the public. There have been calls in America and Europe to regulate Google, YouTube, Facebook, Twitter, and other social-media companies like public utilities. But once again, the pessimists are hyping the problem—and proposing the wrong solution.

Yes, it's possible to retreat into an ideological silo, but scholars who analyze online behavior find that few people actually do that. Most Americans remain moderate in their political views, and public-opinion surveys show that their views haven't changed much in recent decades. The big change—a marked polarization of opinion—has occurred not among the citizenry but among what social scientists call the political class, which encompasses most of the crisis industry: the legislators, political activists, campaign contributors, journalists, lobbyists, and scholars who battle over public policy. They're the ones segregating themselves at opposite ends of the political spectrum and trying to drag the rest of the country along. Their vicious battling has created a widespread sense of "false polarization," as the political scientist Morris Fiorina calls it. The typical Democratic and Republican voters accurately consider themselves moderate, but they inaccurately believe that voters in the other party have become dangerously extreme, so they feel increasingly antagonistic toward the other side.

This growing animosity is often blamed on social media, but the trend was under way well before the rise of Facebook, and the partisan animosity is just as strong, if not stronger, among older Americans even though they spend relatively little time on the Internet. (They spend relatively more time watching cable news.) False polarization isn't driven by social media. No, the blame lies mainly with old-fashioned mass media and politicians and the rest of the crisis industry: the shrill voices constantly scaring the public and demonizing anyone who doesn't agree with their solution to the latest crisis. Watching them go at each other all day long has the same effect as watching crime stories every night. You start to believe the problem is worse than it is.

You can't stop professional fearmongers from doing their job, and you can't stop their tidings from triggering the ancient regions of your brain, but you don't have to watch them in the first place. We've compared the negativity bias to the urge to load up on fattening calories: Both were adaptive on the ancestral savanna but are problematic in today's environment. The Great Enrichment has led to a great expansion of waistlines because there's so much fattening food tempting us all day long. That abundance, combined with the federal government's dubious advice in the 1980s and 1990s to substitute carbohydrates for fat, led to a long sharp climb in the obesity rate for a quarter century. But that trend has moderated in the past fifteen years—the obesity rate has actually declined some years—as many Americans have started changing their diets. They're eating more nuts, whole grains, and whole fruits, and they've cut back on sodas and other sugary drinks. In our book on self-control, *Willpower,* we noted that junk food may well be the most difficult of all temptations to resist, but people are learning to do it.

We hope to see a similar trend in the consumption of information: the low-bad diet. The merchants of bad are no more irresistible than the merchants of junk food. The same basic approaches for dealing with the power of bad in your personal relationships and business—minimize the negative, accentuate the positive—can enable you to overcome the negative bias that skews politics and public opinion. When there's a school shooting or a terrorist attack, don't wallow for hours watching the live coverage. When politicians and pundits are assailing each other, switch channels. If you try to follow the Rule of Four by watching four uplifting stories for every bad one, you'll spend a lot less time on all-news stations. This takes some effort, but it's getting easier now that there are so many new alternatives.

Old media was a tabloid screaming about the latest crime or scandal. New media, as George Gilder predicted, is more like a bookstore with thousands of books appealing to all kinds of nonlurid interests. For all the snark and bile and trolling on social media, it's actually more Pollyannaish than mass media. As we noted in chapter 8, people prefer to post photos of natural wonders rather than massacres. They form online groups based on shared intellectual and cultural passions rather than a shared contempt for political enemies.

They learn about history and cosmology from professors lecturing on YouTube and reap the benefits of nostalgizing with old friends on Instagram or Facebook. They link to books and long stories about scientific progress and cultural innovation. By choosing your online friends carefully and curating your news feed, you'll get a much sunnier—and more accurate—view of the world. You'll see a lot more than four good things for every bad one.

Every new technology brings new problems, but we're confident that the Digital Revolution's benefits will far exceed its costs. In the finest tradition of the Great Enrichment, it is liberating people from centralized authority, freeing them to experiment and share their innovations. There will always be social stupidities and moral panics, but individuals can learn to think for themselves. That's why we wrote this book.

We share the optimism of Charles Mackay, the Scottish journalist who identified the Crisis Crisis long before we did. He didn't use that label, but he clearly recognized the problem in his classic book *Extraordinary Popular Delusions and the Madness of Crowds.* Published in 1841, it recounted the manias that inspired witch hunts, speculative economic bubbles, end-of-the-world panics, and other "popular follies" throughout history. Mackay didn't offer much in the way of diagnosis—this was long before social psychologists and economists were theorizing about negativity bias and social stupidities—and he didn't pretend to offer any grand political solutions. He did not propose to remake society. His goal, like ours, was simply to enlighten each of his readers.

"Men, it has been well said, think in herds," he wrote. "It will be seen that they go mad in herds, while they only recover their senses slowly, and one by one."

We're all subject to the negativity effect. But one by one, each of us can overcome it or turn it to the advantage of ourselves and society. That is how life improves and civilization advances. Bad is stronger, and at times it may seem indomitable, but we're confident that good will prevail.

Acknowledgments

Writing a book inevitably means some difficult moments, but we got so much help from friends and colleagues that the good overwhelmed the bad. We're indebted, first of all, to the psychologists who collaborated with Roy Baumeister on the original "Bad Is Stronger Than Good" paper: Ellen Bratslavsky, Catrin Finkenauer, and Kathleen Vohs. We drew from Martin Seligman's pioneering work on positive psychology and Rodney Stark's revisionist histories of religion and technological innovation. In sorting out the good long-term trends from the bad news of the day, we drew from a distinguished line of optimists starting with Julian Simon and continuing with Gregg Easterbrook (also an expert on the folly of punting on fourth down), Stephen Moore, Matt Ridley, Ronald Bailey, and Steven Pinker. We're indebted to Max Roser and his team at Our World in Data for so carefully plotting the good trends in modern life.

We received generous aid and guidance from researchers and writers in a wide variety of fields, including Brian Anderson, Chris Anderson, Adryenn Ashley, Jesse Ausubel, Jonah Berger, Paul Beston, David

Black (who pointed us to Trollope's exploration of the negativity effect), Lucy Brown, Eliza Byington, Nathan DeWall, Joseph DiMasi, Peter Sheridan Dodds, Chris Emmins, Will Felps, Christopher Ferguson, Eli Finkel, Morris Fiorina, Harris Friedman, Matthew Gentzkow, Daniel Gilbert, George Gilder, Bill Godshall, Dick Grote, Jonathan Haidt, Matt Hutson, Kay Hymowitz, Dominic Johnson, Angela Legg, John List, Heather Mac Donald, Douglas Maynard, Carey Morewedge, John Mueller, Gerry Ohrstrom, P. J. O'Rourke, Jonathan Rauch, Brad Rodu, Constantine Sedikides, Azim Shariff, Michael Siegel, Lenore Skenazy, Jacob Sullum, Cass Sunstein, Robert Sutton, Kate Sweeny, Philip Tetlock, Richard Thaler, and Tevi Troy. Thanks to them all, and to those who shared their stories with us in interviews, including Coach Kevin Kelley of Pulaski Academy; Felix Baumgartner, Michael Gervais, Joe Kittinger, Art Thompson, and the rest of the Red Bull Stratos team; Eva Moskowitz and Ann Powell at Success Academy; and Adele Gutman and John Taboada at the Casablanca Hotel.

We owe special thanks to the John Templeton Foundation (and in particular to Christopher Levenick) for supporting this book project, and for its funding of research by Baumeister and other psychologists. John Tierney's reporting was supported by the *New York Times* and *City Journal*, which published his articles on some of the topics covered in the book (including nostalgia, the Pollyanna principle, Felix Baumgartner's leap, and electronic cigarettes), and by Columbia University's MFA writing program, which provided an internship for Sukriti Yadava to do research for the book. She was wonderfully creative and energetic in applying her economic training and journalistic instincts to study the negativity effect. We're grateful to her and to Patricia O'Toole, who was the director of the internship program. Baumeister's time and effort were supported by Florida State University, the Russell Sage Foundation, and the University of Queensland, and we appreciate the research assistance of Jordan Landers.

From start to finish, we were buoyed by the encouragement of our literary agent extraordinaire, Kristine Dahl. She helped shape the book's ideas, with expert assistance from Caroline Eisenmann and Tamara Kawar at ICM. Our editor, Virginia Smith, adroitly guided us through the book's iterations and introduced some excellent ideas along the way. We thank her, Caroline Sydney, Maureen Clark, and the rest of the team at Penguin Press.

Last, and certainly not least, we're indebted to the family and friends who endured our obsessions and persevered through the manuscripts and the quest to find the right title. Christopher Buckley, Patrick Cooke, and Patrick Tierney were brave first responders. Helen Fisher heroically maintained positive illusions. Luke Tierney kept calm and carried on. Dianne Tice was sweet, stalwart, and steadfast. Thank you, thank you.

Notes

PROLOGUE: The Negativity Effect

5 **took on questions:** For a list of Baumeister's articles and books, see http://www.roybaumeister.com/.

5 **Baumeister and his colleagues:** R. F. Baumeister, E. Bratslavsky, C. Finkenauer, and K. D. Vohs, "Bad Is Stronger Than Good," *Review of General Psychology* 5 (2001): 323–70.

6 **set of experiments:** P. Rozin, L. Millman, and C. Nemeroff, "Operation of the Laws of Sympathetic Magic in Disgust and Other Domains," *Journal of Personality and Social Psychology* 50 (1986): 703–12.

8 **twenty other languages:** P. Rozin, L. Berman, and E. Royzman, "Biases in Use of Positive and Negative Words Across Twenty Natural Languages," *Cognition and Emotion* 24 (2010): 536–48, https://doi.org/10.1080/02699930902793462.

9 **Rozin's paper, coauthored:** P. Rozin and E. B. Royzman, "Negativity Bias, Negativity Dominance, and Contagion," *Personality and Social Psychology Review* 5 (2001): 296–320.

9 **Baumeister's paper was:** Baumeister, Bratslavsky, Finkenauer, and Vohs, "Bad Is Stronger Than Good," 323–70.

10 **at least 80 percent:** National Collaborating Centre for Mental Health (UK), *Post-Traumatic Stress Disorder: The Management of PTSD in*

Adults and Children in Primary and Secondary Care (Leicester, UK: Gaskell, 2005), https://www.ncbi.nlm.nih.gov/books/NBK56506/. See also E. Jones, "Victimhood: A Traumatic History," *Spiked*, September 2017, https://www.spiked-online.com/2017/09/01/a-traumatic-history/.

12 **Terrorism is a creation:** M. Boot, "The Futility of Terrorism," *Wall Street Journal*, April 16, 2013, https://www.wsj.com/articles/SB10001424127887324485004578426520067308336.

12 **crime in America:** J. Gramlich, "5 Facts About Crime in the U.S.," Pew Research Center, January 3, 2019, http://www.pewresearch.org/fact-tank/2018/01/30/5-facts-about-crime-in-the-u-s/.

12 **because they see:** Ipsos MORI, "Why Do People Think There Is More Crime?," January 2007, quoted in M. Roser and M. Nagdy, "Optimism & Pessimism," Our World in Data, 2019, https://ourworldindata.org/optimism-pessimism.

12 **their personal worries:** W. M. Johnston and G. C. L. Davey, "The Psychological Impact of Negative TV News Bulletins: The Catastrophizing of Personal Worries," *British Journal of Psychology* 88 (1997): 85–91, https://onlinelibrary.wiley.com/doi/abs/10.1111/j.2044-8295.1997.tb02622.x.

13 **less than $1:** M. Roser and E. Ortiz-Ospina, "Global Extreme Poverty," Our World in Data, March 27, 2017, https://ourworldindata.org/extreme-poverty.

13 **how to read:** M. Roser and E. Ortiz-Ospina, "Literacy," Our World in Data, September 20, 2018, https://ourworldindata.org/literacy.

13 **below 10 percent:** World Bank, *Poverty and Shared Prosperity 2018: Piecing Together the Poverty Puzzle* (Washington, DC: World Bank, 2018), 23, and UNESCO Institute for Statistics, "Literacy Rates—UNICEF Data," July 2018, https://data.unicef.org/topic/education/literacy/.

13 **prisoners serving time:** C. Sedikides, R. Meek, M. D. Alicke, and S. Taylor, "Behind Bars but Above the Bar: Prisoners Consider Themselves More Prosocial Than Non-prisoners," *British Journal of Social Psychology* 53 (2014): 396–403.

14 **They've recently drawn:** D. P. Johnson and D. R. Tierney, "Bad World: The Negativity Bias in International Relations," *International Security* 43 (Winter 2018–19): 96-140, https://doi.org/10.1162/ISEC_a_00336.

15 **by illegal immigrants:** The rate of crime among illegal immigrants has been estimated to be comparatively low for most crimes (although the estimates are necessarily inexact because of uncertainty over the number of illegal immigrants in the United States). A 2018 study finds their rate of arrests and convictions to be lower overall than the rate among native-born Americans, and specifically for homicide, sex crimes, and larceny; see A. Nowrasteh, "Criminal Immigrants in Texas: Illegal Immigrant Conviction and Arrest Rates for Homicide, Sex Crimes, Larceny, and Other Crimes," *Cato Institute Immigration Research and Policy Brief No. 4*, February 26, 2018. A separate analysis of Texas data compares their rates with those of the rest of the population (including not only native-born Americans but also legal immigrants, whose rate of crime is lower than that of natives); see B. Latzer, "Do Illegal Aliens Have High Crime Rates?," *City Journal*, January 24, 2019. Latzer finds that rates among illegal immigrants are higher for homicide but lower for burglary, drugs, theft, robbery, and weapons offenses.

15 **abductors of children:** L. Skenazy, *Free-Range Kids: How to Raise Safe, Self-Reliant Children (Without Going Nuts with Worry)* (San Francisco: Wiley, 2009), 16; National Safety Council, *Injury Facts Chart*, https://www.nsc.org/work-safety/tools-resources/injury-facts/chart.

15 **sample of preteen children:** J. M. Chua, "No Kidding, One in Three Children Fear Earth Apocalypse," *TreeHugger*, April 20, 2009, https://www.treehugger.com/culture/no-kidding-one-in-three-children-fear-earth-apocalypse.html.

15 **died in their bathtubs:** J. Mueller and M. Stewart, *Chasing Ghosts: The Policing of Terrorism* (New York: Oxford University Press, 2016), 67.

15 **an availability cascade:** T. Kuran and C. R. Sunstein, "Availability Cascades and Risk Regulation," *Stanford Law Review* 51 (1999): 683–768, http://doi.org/10.2307/1229439, and J. Tierney, "In 2008, a 100 Percent Chance of Alarm," *New York Times*, January 1, 2008, https://www.nytimes.com/2008/01/01/science/01tier.html.

15 **40 percent of Americans:** Mueller and Stewart, *Chasing Ghosts*, 7.

16 **as Thomas Hobbes**: *Leviathan* (Mineola, NY: Dover, 2006), 70.

16 **researchers asked adults:** M. I. Norton, L. Anik, L. B. Aknin, and E. W. Dunn, "Is Life Nasty, Brutish, and Short? Philosophies of Life and Well-Being," *Social Psychological and Personality Science* 2 (2011): 570–75.

16 **to overcome aversion:** P. Rozin, "Getting to Like the Burn of Chili Pepper: Biological, Psychological and Cultural Perspectives," in *Chemical Senses,* vol. 2, *Irritation,* ed. B. G. Green, J. R. Mason, and M. R. Kare (New York: Marcel Dekker, 1990), 231–69.

CHAPTER 1: How Bad Is Bad?

22 **Schwartz concluded that:** R. M. Schwartz, C. F. Reynolds III, M. E. Thase, E. Frank, and A. L. Fasiczka, "Optimal and Normal Affect Balance in Psychotherapy of Major Depression: Evaluation of the Balanced States of Mind Model," *Behavioural and Cognitive Psychotherapy* 30 (2002): 439–50.

23 **a marriage's prospects:** F. Fincham, "Marital Conflict: Correlates, Structure, and Context," *Current Directions in Psychological Science* 12 (2003): 23–27, and J. W. Howard and R. M. Dawes, "Linear Prediction of Marital Happiness," *Personality and Social Psychology Bulletin* 2 (1976): 478–80.

23 **psychologist Harris Friedman:** H. L. Friedman, "A Test of the Validity of the Slater-Sussman Measure of Marital Integration" (master's thesis, Emory University, 1971).

23 **John Gottman found:** R. W. Levenson and J. M. Gottman, "Marital Interaction: Physiological Linkage and Affective Exchange," *Journal of Personality and Social Psychology* 45 (1983): 587–97; R. W. Levenson and J. M. Gottman, "Physiological and Affective Predictors of Change in Relationship Satisfaction," *Journal of Personality and Social Psychology* 49 (1985): 85–94; and J. M. Gottman, "The Roles of Conflict Engagement, Escalation, and Avoidance in Marital Interaction: A Longitudinal View of Five Types of Couples," *Journal of Consulting and Clinical Psychology* 61 (1993): 6–15.

24 **Kahneman and Tversky concluded:** D. Kahneman and A. Tversky, "Choices, Values, and Frames," *American Psychologist* 39 (1984): 341–50.

24 **London and Madrid:** P. D. Windschitl and E. U. Weber, "The Interpretation of 'Likely' Depends on the Context, but '70%' Is 70%—Right? The Influence of Associative Processes on Perceived Certainty,"

Journal of Experimental Psychology: Learning, Memory, and Cognition 25 (1999): 1514–33.

24 **The more familiar:** B. Bilgin, "Losses Loom More Likely Than Gains: Propensity to Imagine Losses Increases Their Subjective Probability," *Organizational Behavior and Human Decision Processes* 118 (2012): 203–15.

24 **gamblers' eye movements:** E. Rubaltelli, S. Dickert, and P. Slovic, "Response Mode, Compatibility, and Dual-Processes in the Evaluation of Simple Gambles: An Eye-Tracking Investigation," *Judgment and Decision Making* 7 (2012): 427–40.

25 **economist Richard Thaler:** R. H. Thaler, "The Value of Saving a Life: A Market Estimate" (Ph.D. diss., University of Rochester, 1974).

25 **tracking workers' moods:** A. G. Miner, T. M. Glomb, and C. Hulin, "Experience Sampling Mood and Its Correlates at Work," *Journal of Occupational and Organizational Psychology* 78 (2005): 171–93.

25 **psychologist Barbara Fredrickson:** B. L. Fredrickson, "Positive Emotions Broaden and Build," *Advances in Experimental Social Psychology* 47 (2013): 1-53.

26 **method for gauging:** Unfortunately, another researcher took the results from Fredrickson's diary research to a ridiculous extreme. The psychologist Marcial Losada analyzed the diary data with an elaborate mathematical model drawn from fluid-dynamic equations. His calculations identified a ratio of exactly 2.9013, which was supposedly the minimum ratio of positive-to-negative emotions that a person needed in order to flourish. The math was beyond most psychologists, but the results looked impressive when they were published in 2005 by a prestigious journal; see B. L. Fredrickson and M. F. Losada, "Positive Affect and the Complex Dynamics of Human Flourishing," *American Psychologist* 60 (2005): 678–86. The 2.9013 positivity ratio became known as the Losada Line and was reverently cited until other researchers concluded that it was nonsense based on arbitrary assumptions and elementary mathematical errors; see N. J. L. Brown, A. D. Sokal, and H. L. Friedman, "The Complex Dynamics of Wishful Thinking: The Critical Positivity Ratio," *American Psychologist* 68 (2013): 801–13. The critique was devastating, and Losada didn't bother responding, while Fredrickson conceded that the complex mathematical analysis was flawed. As a

result, the *American Psychologist* journal formally withdrew the sections of the article involving Losada's calculations while leaving intact Fredrickson's contributions. As embarrassing as the episode was—how could so many social psychologists have been taken in by numerical mumbo-jumbo?—it didn't invalidate Fredrickson's field research or the many other studies that used basic arithmetic to compare the impact of good and bad events, as Fredrickson noted in her 2013 response; see B. L. Fredrickson, "Updated Thinking on Positivity Ratios," *American Psychologist* 68 (2013): 814–22.

27 **psychologist Randy Larsen:** R. J. Larsen, "Differential Contributions of Positive and Negative Affect to Subjective Well-Being," in *Eighteenth Annual Meeting of the International Society for Psychophysics,* ed. J. A. Da Silva, E. H. Matsushima, and N. P. Riberio-Filho (Rio de Janeiro, Brazil: International Society for Psychophysics, 2002), 186–90. See also R. J. Larsen and Z. Prizmic, "Regulation of Emotional Well-Being: Overcoming the Hedonic Treadmill," in *The Science of Subjective Well-Being,* ed. M. Eid and R. J. Larsen (New York: Guilford Press, 2008), 258–89.

29 **assessed in surveys:** D. DiMeglio, "Customer Satisfaction Stagnates in Ominous Sign for Economy, ACSI Data Show," press release, National ACSI Q1 2018, July 12, 2018, http://www.theacsi.org/news-and-resources/press-releases/press-2018/press-release-national-acsi-q1-2018; R. East, K. Hammond, and M. Wright, "The Relative Incidence of Positive and Negative Word of Mouth: A Multi-category Study," *International Journal of Research in Marketing* 24 (2007): 175–84; and T. O. Jones and W. E. Sasser, "Why Satisfied Customers Defect," *Harvard Business Review* (November—December 1995), https://hbr.org/1995/11/why-satisfied-customers-defect.

29 **businesses on Yelp:** https://en.yelp.my/faq#rating_distribution.

29 **Cadbury chocolate company's:** B. Puri and S. E. Clark, "How to Transform Consumer Opinion When Disaster Strikes," *Reflections from Practice* (Medford, MA: Fletcher School, April 2012).

30 **folkore and mythology:** J. Campbell, *The Hero with a Thousand Faces* (New York: Pantheon, 1949).

30 **to external forces:** C. K. Morewedge, "Negativity Bias in Attribution of External Agency," *Journal of Experimental Psychology: General* 138

(2009): 535–45; J. K. Hamlin and A. S. Baron, "Agency Attribution in Infancy: Evidence for a Negativity Bias," *PLoS ONE* 9 (2014): e96112, https://doi.org/10.1371/journal.pone.0096112; and G. Bohner, H. Bless, N. Schwarz, and F. Strack, "What Triggers Causal Attributions? The Impact of Valence and Subjective Probability," *European Journal of Social Psychology* 18 (1988): 335–45. See also M. Hutson, *The 7 Laws of Magical Thinking: How Irrational Beliefs Keep Us Happy, Healthy, and Sane* (New York: Hudson Street Press, 2012).

31 **neighborhoods are safer:** M. Friedman, A. C. Grawert, and J. Cullen, "Crime Trends: 1990–2016," Brennan Center for Justice, April 18, 2017, https://www.brennancenter.org/publication/crime-trends1990-2016.

31 **nation's manufacturing output:** D. Desilver, "Most Americans Unaware That as U.S. Manufacturing Jobs Have Disappeared, Output Has Grown," Pew Research Center, July 25, 2017, http://www.pewresearch.org/fact-tank/2017/07/25/most-americans-unaware-that-as-u-s-manufacturing-jobs-have-disappeared-output-has-grown/.

33 **inefficiency and incompetence:** N. Gillespie, "TSA Celebrates 10 Years of Sucking!," *Reason,* November 17, 2011, http://reason.com/blog/2011/11/17/tsa-10-years-of-sucking.

33 **repeatedly flunked tests:** A. Halsey III, "GOP Report: TSA Hasn't Improved Aviation Security," *Washington Post,* November 16, 2011, https://www.washingtonpost.com/local/commuting/gop-report-tsa-hasnt-improved-aviation-security/2011/11/16/gIQAvqRQSN_story.html?utm_term=.3b3c0c70233b, and J. Tierney, "Monuments to Idiocy: Let's Honor the Public Servants Responsible for Giving Us the TSA," *City Journal,* Summer 2016, https://www.city-journal.org/html/monuments-idiocy-14620.html.

33 **folly was obvious:** J. Tierney, "The Big City: Twisted Logic on Improving Air Security," *New York Times,* October 2, 2001, https://www.nytimes.com/2001/10/02/nyregion/the-big-city-twisted-logic-on-improving-air-security.html, and J. Tierney, "Fighting the Last Hijackers," *New York Times,* August 16, 2005.

33 **additional 1,600 deaths:** W. Gaissmaier and G. Gigerenzer, "9/11, Act II: A Fine-Grained Analysis of Regional Variations in Traffic Fatalities in the Aftermath of the Terrorist Attacks," *Psychological Science* 23 (2012): 1449–54.

34 **Woods once explained:** Associated Press, "Tiger Grabs Doral Lead," *Gainesville Sun,* March 24, 2007, https://www.gainesville.com/article/LK/20070324/News/604156668/GS/.

34 **the economists calculate:** D. G. Pope and M. E. Schweitzer, "Is Tiger Woods Loss Averse? Persistent Bias in the Face of Experience, Competition, and High Stakes," *American Economic Review* 101 (2011): 129–57.

35 **a bad strategy:** D. Romer, "Do Firms Maximize? Evidence from Professional Football," *Journal of Political Economy* 114 (2006): 340–65, and B. Cohen, "NFL Teams Should (Almost) Always 'Go for It' on 4th and 1," Eagles Rewind, August 13, 2013, https://eaglesrewind.com/2013/08/13/nfl-teams-should-almost-always-go-for-it-on-4th-and-1/.

35 **Gregg Easterbrook, the Tuesday Morning Quarterback:** G. Easterbrook, *The Game's Not Over* (New York: PublicAffairs, 2015), 173-82.

35 ***New York Times'* Upshot:** "4th Down: When to Go for It and Why," *New York Times,* September 4, 2014, https://www.nytimes.com/2014/09/05/upshot/4th-down-when-to-go-for-it-and-why.html.

35 **denounced by sportscasters:** Associated Press, "Belichick Defends Decision to Go for It on Fourth Down vs. Colts," NFL.com, November 16, 2009, http://www.nfl.com/news/story/09000d5d81441ff6/article/belichick-defends-decision-to-go-for-it-on-fourth-down-vs-colts.

36 **coach who never punts:** Quotations and facts in this section are drawn from interviews with Kevin Kelley and from his team's records.

38 **experiments with gambling:** E. Polman, "Self-Other Decision Making and Loss Aversion," *Organizational Behavior and Human Decision Processes* 119 (2012): 141–50.

41 **Coach of the Year:** J. Halley, "All-USA Football Coach of the Year: Kevin Kelley, Pulaski Academy (Ark.)," *USA Today,* December 20, 2016.

CHAPTER 2: Love Lessons

44 **ratings typically go downhill:** J. VanLaningham, D. R. Johnson, and P. Amato, "Marital Happiness, Marital Duration, and the U-Shaped Curve: Evidence from a Five-Wave Panel Study," *Social Forces* 79 (2001): 1313–41.

44 **shock to Anthony Trollope:** A. Trollope, *He Knew He Was Right* (London: Ward, Lock, 1869).

46 **a classic study:** C. E. Rusbult, D. J. Johnson, and G. D. Morrow, "Impact of Couple Patterns of Problem Solving on Distress and Nondistress in Dating Relationships," *Journal of Personality and Social Psychology* 50 (1986): 744–53.

47 **your loyalty often:** S. M. Drigotas, G. A. Whitney, and C. E. Rusbult, "On the Peculiarities of Loyalty: A Diary Study of Responses to Dissatisfaction in Everyday Life," *Personality and Social Psychology Bulletin* 21 (1995): 596–609, and N. C. Overall, C. G. Sibley, and L. K. Travaglia, "Loyal but Ignored: The Benefits and Costs of Constructive Communication Behavior," *Personal Relationships* 17 (2010): 127–48.

47 **said Caryl Rusbult:** Personal communication.

47 **project called PAIR:** T. L. Huston, J. P. Caughlin, R. M. Houts, S. E. Smith, and L. J. George, "The Connubial Crucible: Newlywed Years as Predictors of Marital Delight, Distress, and Divorce," *Journal of Personality and Social Psychology* 80 (2001): 237–52.

48 **test a theory:** S. L. Murray, P. Rose, G. M. Bellavia, J. G. Holmes, and A. G. Kusche, "When Rejection Stings: How Self-Esteem Constrains Relationship-Enhancement Processes," *Journal of Personality and Social Psychology* 83 (2002): 556–73, and S. L. Murray, J. G. Holmes, and N. L. Collins, "Optimizing Assurance: The Risk Regulation System in Relationships," *Psychological Bulletin* 132 (2006): 641–66.

49 **Murray and Holmes found:** S. L. Murray et al., "Cautious to a Fault: Self-Protection and the Trajectory of Marital Satisfaction," *Journal of Experimental Social Psychology* 49 (2013): 522–33, https://doi.org/10.1016/j.jesp.2012.10.010.

50 **New York City couples:** G. Downey, A. L. Freitas, B. Michaelis, and H. Khouri, "The Self-Fulfilling Prophecy in Close Relationships: Rejection Sensitivity and Rejection by Romantic Partners," *Journal of Personality and Social Psychology* 75 (1998): 545–60.

51 **couples in Seattle:** R. W. Levenson and J. M. Gottman, "Marital Interaction: Physiological Linkage and Affective Exchange," *Journal of Personality and Social Psychology* 45 (1983): 587–97, and R. W. Levenson and J. M. Gottman, "Physiological and Affective Predictors of Change

in Relationship Satisfaction," *Journal of Personality and Social Psychology* 49 (1985): 85–94.

52 **group of same-sex couples:** J. M. Gottman et al., "Observing Gay, Lesbian and Heterosexual Couples' Relationships: Mathematical Modeling of Conflict Interaction," *Journal of Homosexuality* 45 (2003): 65–91.

52 **communication between newlyweds:** T. L. Huston and A. L. Vangelisti, "Socioemotional Behavior and Satisfaction in Marital Relationships: A Longitudinal Study," *Journal of Personality and Social Psychology* 61 (1991): 721–33.

52 **psychologist Barry McCarthy:** Personal communication with B. McCarthy.

53 **the "propinquity effect":** L. Festinger, S. Schachter, and K. Back, *Social Pressures in Informal Groups: A Study of Human Factors in Housing* (Stanford, CA: Stanford University Press, 1950).

53 **a town-house development:** E. B. Ebbesen, G. L. Kjos, and V. J. Konečni, "Spatial Ecology: Its Effects on the Choice of Friends and Enemies," *Journal of Experimental Social Psychology* 12 (1976): 505–18.

54 **children's cognitive development:** D. C. Rowe, K. C. Jacobson, and E. J. Van den Oord, "Genetic and Environmental Influences on Vocabulary IQ: Parental Education Level as Moderator," *Child Development* 70 (1999): 1151–62; L. A. Thompson, R. D. Tiu, and D. K. Detterman, "Differences in Heritability Across Levels of Father's Occupation," poster presented at the annual meeting of the Behavior Genetics Association (July 1999); and E. Turkheimer, A. Haley, M. Waldron, B. D'Onofrio, and I. I. Gottesman, "Socioeconomic Status Modifies Heritability of IQ in Young Children," *Psychological Science* 14 (2003): 623–28.

55 **emotional aspects of child rearing:** S. Scarr, "Developmental Theories for the 1990s: Development and Individual Differences," *Child Development* 63 (1992): 1–19, and I. B. Wissink, M. Dekovic, and A. M. Meijer, "Parenting Behavior, Quality of the Parent-Adolescent Relationship, and Adolescent Functioning in Four Ethnic Groups," *Journal of Early Adolescence* 26 (2006): 133–59.

56 **Ayelet Gneezy demonstrated:** A. Gneezy and N. Epley, "Worth Keeping but Not Exceeding: Asymmetric Consequences of Breaking Versus Exceeding Promises," *Social Psychological and Personality Science* 5 (2014): 796–804.

57 **dozens of countries:** D. P. Schmitt, A. Realo, M. Voracek, and J. Allik, "Why Can't a Man Be More Like a Woman? Sex Differences in Big Five Personality Traits Across 55 Cultures," *Journal of Personality and Social Psychology* 94 (2008): 168–82.

57 **are shown faces:** L. M. Williams et al., "Explicit Identification and Implicit Recognition of Facial Emotions: I. Age Effects in Males and Females Across 10 Decades," *Journal of Clinical and Experimental Neuropsychology* 31 (2009): 257–77.

59 ***fundamental attribution error:*** D. T. Gilbert and P. S. Malone, "The Correspondence Bias," *Psychological Bulletin* 117 (1995): 21–38.

59 **error occurs only:** B. F. Malle, "The Actor-Observer Asymmetry in Attribution: A (Surprising) Meta-analysis," *Psychological Bulletin* 132 (2006): 895–919.

60 **their "attributional style":** B. R. Karney and T. N. Bradbury, "Attributions in Marriage: State or Trait? A Growth Curve Analysis," *Journal of Personality and Social Psychology* 78 (2000): 295–309.

60 **the "marriage hack":** E. J. Finkel, E. B. Slotter, L. B. Luchies, G. M. Walton, and J. J. Gross, "A Brief Intervention to Promote Conflict Reappraisal Preserves Marital Quality over Time," *Psychological Science* 24 (2013): 1595–601.

61 **brains of lovers:** X. Xu et al., "Regional Brain Activity During Early-Stage Intense Romantic Love Predicted Relationship Outcomes After 40 Months: An fMRI Assessment," *Neuroscience Letters* 526 (2012): 33–38, and B. P. Acevedo, A. Aron, H. E. Fisher, and L. L. Brown, "Neural Correlates of Marital Satisfaction and Well-Being: Reward, Empathy, and Affect," *Clinical Neuropsychiatry: Journal of Treatment Evaluation* 9 (2012): 20–31.

62 **poet William Blake:** W. Blake, *The Poetical Works of William Blake*, ed. John Sampson (London: Oxford University Press, 1908); https://www.bartleby.com/235/154.html.

62 **illusions of couples:** S. L. Murray, J. G. Holmes, and D. W. Griffin, "The Benefits of Positive Illusions: Idealization and the Construction of Satisfaction in Close Relationships," *Journal of Personality and Social Psychology* 70 (1996): 79–98.

63 **Ginsburg passed that advice:** K. J. Sullivan, "U.S. Supreme Court Justice Ruth Bader Ginsburg Talks About a Meaningful Life," *Stanford News,* February 6, 2017, https://news.stanford.edu/2017/02/06/supreme-court-associate-justice-ginsburg-talks-meaningful-life/.

64 **game called Dictator:** B. Keysar, B. A. Converse, J. Wang, and N. Epley, "Reciprocity Is Not Give and Take: Asymmetric Reciprocity to Positive and Negative Acts," *Psychological Science* 19 (2008): 1280–86, https://dx.doi.org/10.1111/j.1467-9280.2008.02223.x.

CHAPTER 3: The Brain's Inner Demon

67 **Felix Baumgartner aspired:** The account of the supersonic-jump project draws from interviews with Felix Baumgartner, Mike Gervais, Art Thompson, Joe Kittinger, and other members of the Red Bull Stratos Team, as well as Tierney's other reporting during the project. See J. Tierney, "Daredevil Sets Sight on a 22-Mile Fall," *New York Times,* October 8, 2012, https://www.nytimes.com/2012/10/09/science/fearless-felix-baumgartner-to-try-to-become-first-sky-diver-to-break-sound-barrier.html, and J. Tierney, "A Supersonic Jump, from 23 Miles in the Air," *New York Times,* March 15, 2010, https://www.nytimes.com/2010/03/16/science/16tier.html.

70 **major threat-warning systems:** K. J. Flannelly, H. G. Koenig, K. Galek, and C. G. Ellison, "Beliefs, Mental Health, and Evolutionary Threat Assessment Systems in the Brain," *Journal of Nervous and Mental Disease* 195 (2007): 996–1003.

71 **rest of the brain:** M. Diano, A. Celeghin, A. Bagnis, and M. Tamietto, "Amygdala Response to Emotional Stimuli Without Awareness: Facts and Interpretations," *Frontiers in Psychology* 7 (2017): 2029, https://doi.org/10.3389/fpsyg.2016.02029.

71 **turn more quickly:** V. LoBue and J. S. DeLoache, "Superior Detection of Threat-Relevant Stimuli in Infancy," *Developmental Science* 13 (2010): 221–28, https://doi.org/10.1111/j.1467-7687.2009.00872.x.

71 **By age five:** V. LoBue, "More Than Just Another Face in the Crowd: Superior Detection of Threatening Facial Expressions in Young

Children and Adults," *Developmental Science* 12 (2009): 305–13, https://doi.org/10.1111/j.1467-7687.2008.00767.x.

71 **brain focuses longer:** G. W. Alpers and P. Pauli, "Emotional Pictures Predominate in Binocular Rivalry," *Cognition and Emotion* 20 (2006): 596–607, and R. L. Bannerman, M. Milders, B. de Gelder, and A. Sahraie, "Influence of Emotional Facial Expressions on Binocular Rivalry," *Ophthalmic & Physiological Optics* 28 (2008): 317–26.

72 **one goggle experiment:** E. Anderson, E. H. Siegel, E. Bliss-Moreau, and L. F. Barrett, "The Visual Impact of Gossip," *Science* 332 (2011): 1446–48.

72 **the Stroop test:** F. Pratto and O. P. John, "Automatic Vigilance: The Attention-Grabbing Power of Negative Social Information," *Journal of Personality and Social Psychology* 61 (1991): 380–91.

72 **reflexes to withdraw:** G. J. Norman et al., "Current Emotion Research in Psychophysiology: The Neurobiology of Evaluative Bivalence," *Emotion Review* 3 (2011): 349–59, https://doi.org/10.1177/1754073911402403.

73 **To test its flexibility:** W. A. Cunningham, J. J. Van Bavel, and I. Johnsen Haas, "Affective Flexibility: Evaluative Processing Goals Shape Amygdala Activity," *Psychological Science* 19 (2008): 152–60.

73 **amygdala keeps looking:** M. W. Schlund and M. F. Cataldo, "Amygdala Involvement in Human Avoidance, Escape and Approach Behavior," *Neuroimage* 53 (2010): 769–76.

74 **damage in the amygdala:** J. A. Weller, I. P. Levin, B. Shiv, and A. Bechara, "Neural Correlates of Adaptive Decision Making for Risky Gains and Losses," *Psychological Science* 18 (2007): 958–64.

75 **two playing cards:** T. D. Satterthwaite et al., "Dissociable but Interrelated Systems of Cognitive Control and Reward During Decision Making: Evidence from Pupillometry and Event-Related fMRI," *Neuroimage* 37 (2007): 1017–31.

75 **not consciously fearful:** G. Hochman and E. Yechiam, "Loss Aversion in the Eye and in the Heart: The Autonomic Nervous System's Responses to Losses," *Journal of Behavioral Decision Making* 24 (2011): 140–56.

78 **formed "positive illusions":** S. E. Taylor, M. E. Kemeny, G. M. Reed, J. E. Bower, and T. L. Gruenewald, "Psychological Resources, Positive

Illusions, and Health," *American Psychologist* 55 (2000): 99–109, and S. E. Taylor, "Adjustment to Threatening Events: A Theory of Cognitive Adaptation," *American Psychologist* 38 (1983): 1161–73.

78 **mobilization and minimization:** S. E. Taylor, "Asymmetrical Effects of Positive and Negative Events: The Mobilization-Minimization Hypothesis," *Psychological Bulletin* 110 (1991): 67–85.

81 **in treating depression:** R. J. DeRubeis, G. J. Siegle, and S. D. Hollon, "Cognitive Therapy Versus Medication for Depression: Treatment Outcomes and Neural Mechanisms," *Nature Reviews Neuroscience* 9 (2008): 788–96.

81 **studied so thoroughly:** A. C. Butler, J. E. Chapman, E. M. Forman, and A. T. Beck, "The Empirical Status of Cognitive-Behavioral Therapy: A Review of Meta-analyses," *Clinical Psychology Review* 26 (2006): 17–31, and B. O. Olatunji, J. M. Cisler, and B. J. Deacon, "Efficacy of Cognitive Behavioral Therapy for Anxiety Disorders: A Review of Meta-analytic Findings," *Psychiatric Clinics of North America* 33 (2010): 557–77.

81 **range of problems:** Academy of Cognitive Therapy, "CBT Outcome Studies," https://www.academyofct.org/page/OutcomeStudies.

81 **depression, anxiety, and other disorders:** C. Otte, "Cognitive Behavioral Therapy in Anxiety Disorders: Current State of the Evidence," *Dialogues in Clinical Neuroscience* 13 (2011): 413–21, and C. K. Higa-McMillan, S. E. Francis, L. Rith-Najarian, and B. F. Chorpita, "Evidence Base Update: 50 Years of Research on Treatment for Child and Adolescent Anxiety," *Journal of Clinical Child & Adolescent Psychology* 45 (2016): 91–113.

82 **the "coping statement":** D. Roditi, M. E. Robinson, and N. Litwins, "Effects of Coping Statements on Experimental Pain in Chronic Pain Patients," *Journal of Pain Research* 2 (2009): 109–16, and F. D. Glogower, W. J. Fremouw, and J. C. McCroskey, "A Component Analysis of Cognitive Restructuring," *Cognitive Therapy and Research* 2 (1978): 209–23.

82 **the relaxation response:** H. Benson and M. Z. Klipper, *The Relaxation Response* (New York: William Morrow, 1975).

84 **to sharing information:** W. Walker, J. J. Skowronski, J. A. Gibbons, R. J. Vogl, and T. D. Ritchie, "Why People Rehearse Their Memories:

Frequency of Use and Relations to the Intensity of Emotions Associated with Autobiographical Memories," *Memory* 17 (2009): 760–73.

85 **monks and practitioners:** R. Jevning, R. K. Wallace, and M. Beidebach, "The Physiology of Meditation: A Review," *Neuroscience & Biobehavioral Reviews* 16 (1992): 415–24, and A. Hankey, "Studies of Advanced Stages of Meditation in the Tibetan Buddhist and Vedic Traditions," *Evidence-Based Complementary and Alternative Medicine* 3 (2006): 513–21.

85 **blood pressure drops:** S. L. Ooi, M. Giovino, and S. C. Pak, "Transcendental Meditation for Lowering Blood Pressure: An Overview of Systematic Reviews and Meta-analyses," *Complementary Therapies in Medicine* 34 (2017): 26–34.

86 **great mass potential:** B. Carey, "The Therapist May See You Anytime, Anywhere," *New York Times,* February 13, 2012, https://www.nytimes.com/2012/02/14/health/feeling-anxious-soon-there-will-be-an-app-for-that.html?_r=0.

86 **a smartphone app:** P. M. Enock, S. G. Hofmann, and R. J. McNally, "Attention Bias Modification Training via Smartphone to Reduce Social Anxiety: A Randomized, Controlled Multi-session Experiment," *Cognitive Therapy and Research* 38 (2014): 200–216, and R. Yang et al., "Effects of Cognitive Bias Modification Training via Smartphones," *Frontiers in Psychology* 8 (2017): 1370.

86 **to lower anxiety:** J. M. Kuckertz and N. Amir, "Attention Bias Modification for Anxiety and Phobias: Current Status and Future Directions," *Current Psychiatry Reports* 17 (2015): 1–8; C. Mogoașe, D. David, and E. H. Koster, "Clinical Efficacy of Attentional Bias Modification Procedures: An Updated Meta-analysis," *Journal of Clinical Psychology* 70 (2014): 1133–57; and L. S. Hallion and A. M. Ruscio, "A Meta-analysis of the Effect of Cognitive Bias Modification on Anxiety and Depression," *Psychological Bulletin* 137 (2011): 940–58.

86 **giving a speech:** N. Amir, G. Weber, C. Beard, J. Bomyea, and C. T. Taylor, "The Effect of a Single-Session Attention Modification Program on Response to a Public-Speaking Challenge in Socially Anxious Individuals," *Journal of Abnormal Psychology* 117 (2008): 860–68.

86 **fear of heights:** S. A. Steinman and B. A. Teachman, "Reaching New Heights: Comparing Interpretation Bias Modification to Exposure

Therapy for Extreme Height Fear," *Journal of Consulting and Clinical Psychology* 82 (2014): 404–17.

90 **setting a record:** J. Tierney, "24 Miles, 4 Minutes and 834 M.P.H., All in One Jump," *New York Times,* October 14, 2012, https://www.nytimes.com/2012/10/15/us/felix-baumgartner-skydiving.html.

CHAPTER 4: Use the Force

91 **with Stephen Potter:** S. Potter, *The Complete Upmanship, Including Gamesmanship, Lifemanship, One-Upmanship and Supermanship* (New York: Holt, Rinehart and Winston, 1971).

92 **some clever experiments:** T. M. Amabile, "Brilliant but Cruel: Perceptions of Negative Evaluators," *Journal of Experimental Social Psychology* 19 (1983): 146–56.

94 **demonstrated in another study:** T. M. Amabile and A. H. Glazebrook, "A Negativity Bias in Interpersonal Evaluation," *Journal of Experimental Social Psychology* 18 (1982): 1–22.

94 **by Elizabeth Bennet:** J. Austen, *Pride and Prejudice* (New York: Penguin Classics, 2009), 196–97, Kindle.

95 **To test these phrases:** R. Hamilton, K. D. Vohs, and A. L. McGill, "We'll Be Honest, This Won't Be the Best Article You'll Ever Read: The Use of Dispreferred Markers in Word-of-Mouth Communication," *Journal of Consumer Research* 41 (2014): 197–212.

96 **Mary Kay Ash:** M. K. Ash, *Mary Kay on People Management* (London: Futura, 1984), 39.

96 **When Douglas Maynard:** D. W. Maynard, *Bad News, Good News: Conversational Order in Everyday Talk and Clinical Settings* (Chicago: University of Chicago Press, 2003).

97 **More than three-quarters:** A. M. Legg and K. Sweeny, "Do You Want the Good News or the Bad News First? The Nature and Consequences of News Order Preferences," *Personality and Social Psychology Bulletin* 40 (2014): 279–88, https://doi.org/10.1177/0146167213509113, and L. L. Marshall and R. F. Kidd, "Good News or Bad News First?," *Social Behavior and Personality* 9 (1981): 223–26.

97 **as Baumeister found:** R. F. Baumeister and K. J. Cairns, "Repression and Self-Presentation: When Audiences Interfere with Self-Deceptive Strategies," *Journal of Personality and Social Psychology* 62 (1992): 851–62.

98 **Thomas Jefferson's observation:** T. Jefferson, "Letter to Francis Hopkinson, March 13, 1789," Founders Online, https://founders.archives.gov/documents/Jefferson/01-14-02-0402.

100 **"perspective display sequence":** Drawn from personal communication with Douglas Maynard and from his book *Bad News, Good News.*

101 **person will feel better:** A. Nguyen, A. M. Legg, and K. Sweeny, "Do You Want the Good News or the Bad News First? News Order Influences Recipients' Mood, Perceptions, and Behaviors," *University of California Riverside Undergraduate Research Journal* 5 (2011): 31–36.

101 **study of sequencing strategies:** Legg and Sweeny, "Do You Want the Good News or the Bad News First?"

102 **subsequent negative feedback:** Y. Trope and E. Neter, "Reconciling Competing Motives in Self-Evaluation: The Role of Self-Control in Feedback Seeking," *Journal of Personality and Social Psychology* 66 (1994): 646–57, https://dx.doi.org/10.1037/0022-3514.66.4.646.

103 ***In doling out praise:*** E. Chan and J. Sengupta, "Insincere Flattery Actually Works: A Dual Attitudes Perspective," *Journal of Marketing Research* 47 (2010): 122–33; C. Nass, *The Man Who Lied to His Laptop* (New York: Current, 2010), 16–38.

104 **When Ronald Reagan:** E. I. Koch, "Reagan's Afterlife on Earth," *Jewish World Review,* June 8, 2004, http://jewishworldreview.com/0604/koch_reagan.php3.

105 **Lee Daniels has:** S. W. Hunt and L. Rose, "Lee Daniels, Damon Lindelof, A-List Writers on Race, Ignoring Critics, an 'Empire' Axing," *Hollywood Reporter,* May 11, 2015, https://www.hollywoodreporter.com/features/lee-daniels-damon-lindelof-a-794430.

105 **Winston Churchill observed:** Parliamentary Debates (Hansard), January 27, 1940, quoted in W. S. Churchill and R. Langworth, *Churchill by Himself: In His Own Words* (New York: RosettaBooks, 2013), "Maxims," Kindle.

CHAPTER 5: Heaven or Hell

107 **Lexicologists have traced:** M. Quinion, "Carrot and Stick," World Wide Words, 2009, http://www.worldwidewords.org/qa/qa-car4.htm, and J. Freeman, "Carrot Unstuck: A New Twist in an Old Debate," Boston.com, March 8, 2009, http://archive.boston.com/bostonglobe/ideas/articles/2009/03/08/carrot_unstuck/.

107 **widely reprinted story:** "The Neighbour-in-Law," in L. M. Child, *Fact and Fiction: A Collection of Stories* (New York: C. S. Francis, 1846), 156–57.

107 **an American magazine:** *The Eclectic Magazine,* August 1851, quoted in Quinion, "Carrot and Stick."

108 **coax their mule teams:** G. C. Armistead, *Horses and Mules in the Civil War: A Complete History with a Roster of More Than 700 War Horses* (Jefferson, NC: McFarland, 2013).

108 **an impious people:** R. Finke and R. Stark, *The Churching of America, 1776–2005: Winners and Losers in Our Religious Economy* (New Brunswick, NJ: Rutgers University Press, 2005), chap. 2, Kindle.

108 **downed six drinks:** J. Kross, "'If You Will Not Drink with Me, You Must Fight with Me': The Sociology of Drinking in the Middle Colonies," *Pennsylvania History* 64 (1997): 28–55.

109 **Benjamin Franklin reported:** B. Franklin, *Autobiography of Benjamin Franklin* (New York: Henry Holt, 1916), chap. XI, Kindle.

109 **Hell on Earth:** Historical material in this section is drawn from Finke and Starke, *Churching of America.*

110 **sermon in Georgia:** G. Whitefield, *The Collected Sermons of George Whitefield* (Jawbone Digital, 2015), loc. 7265–7304 of 14,042, Kindle.

110 **sermon of 1741:** J. Edwards, *Sinners in the Hands of an Angry God* (Musaicum Books, 2018), "1. Sinners in the Hands of an Angry God," Kindle.

112 **Bishop Wilke explained:** R. B. Wilke, *And Are We Yet Alive?* (Nashville: Abingdon Press, 1986), 98.

113 **conceived of God:** A. F. Shariff and A. Norenzayan, "Mean Gods Make Good People: Different Views of God Predict Cheating

Behavior," *International Journal for the Psychology of Religion* 21 (2011): 85–96. See also O. Yilmaz and H. G. Bahçekapili, "Supernatural and Secular Monitors Promote Human Cooperation Only If They Remind of Punishment," *Evolution and Human Behavior* 37 (2016): 79–84.

113 **compared crime rates:** A. F. Shariff and M. Rhemtulla, "Divergent Effects of Beliefs in Heaven and Hell on National Crime Rates," *PLoS ONE* 7 (2012): e39048. https://doi.org/10.1371/journal.pone.0039048. See also "Updated Heaven, Hell and Crime Data for Shariff & Rhemtulla 2012," Sharifflab, September 8, 2015, http://sharifflab.com/updated-heaven-hell-and-crime-data-for-shariff-rhemtulla-2012/.

114 **John Garcia reported:** J. Garcia, D. J. Kimeldorf, and R. A. Koelling, "Conditioned Aversion to Saccharin Resulting from Exposure to Gamma Radiation," *Science* 122 (1955): 157–58.

115 **Researchers in Belgium:** F. Baeyens, P. Eelen, O. Van den Bergh, and G. Crombez, "Flavor-Flavor and Color-Color Conditioning in Humans," *Learning and Motivation* 21 (1990): 434–55.

115 **children and marbles:** A. F. Costantini and K. L. Hoving, "The Effectiveness of Reward and Punishment Contingencies on Response Inhibition," *Journal of Experimental Child Psychology* 16 (1973): 484–94.

116 **Your memory improves:** J. P. Forgas, "Don't Worry, Be Sad! On the Cognitive, Motivational, and Interpersonal Benefits of Negative Mood," *Current Directions in Psychological Science* 22 (2013): 225–32.

116 **also more succinct:** A. S. Koch, J. P. Forgas, and D. Matovic, "Can Negative Mood Improve Your Conversation? Affective Influences on Conforming to Grice's Communication Norms," *European Journal of Social Psychology* 43 (2013): 326–34.

116 **were less biased:** J. P. Forgas, "Can Negative Affect Eliminate the Power of First Impressions? Affective Influences on Primacy and Recency Effects in Impression Formation," *Journal of Experimental Social Psychology* 41 (2011): 425–29.

116 **ability to spot liars:** J. P. Forgas and R. East, "On Being Happy and Gullible: Mood Effects on Skepticism and the Detection of Deception," *Journal of Experimental Social Psychology* 44 (2008): 1362–67.

116 **blasts of noise:** L. Notebaert, M. Tilbrook, P. J. F. Clarke, and C. Macleod, "When a Bad Bias Can Be Good: Anxiety-Linked Attentional Bias to Threat in Contexts Where Dangers Can Be Avoided," *Clinical Psychological Science* 5 (2017): 485–96.

117 **fear of looking fat:** S. E. Dalley. P. Toffanin, and T. V. Pollet, "Dietary Restraint in College Women: Fear of an Imperfect Fat Self Is Stronger Than Hope of a Perfect Thin Self," *Body Image* 9 (2012): 441–47.

117 **for blood donations:** E. Y. Chou and J. K. Murnighan, "Life or Death Decisions: Framing the Call for Help," *PLoS ONE* 8 (2013): e57351, https://doi.org/10.1371/journal.pone.0057351.

117 **Dozens of other studies:** M. B. Tannenbaum et al., "Appealing to Fear: A Meta-analysis of Fear Appeal Effectiveness and Theories," *Psychological Bulletin* 141 (2015): 1178–204.

117 **Dr. Johnson observed:** "Depend upon it, Sir, when a man knows he is to be hanged in a fortnight, it concentrates his mind wonderfully." Quoted in J. Boswell, *The Life of Samuel Johnson,* C. Hibbert, ed. (London: Penguin English Library, 1979), 231.

117 **parents and educators:** R. Ryan, A. Kilal, K. Ziol-Guest, and C. Padilla, "Socioeconomic Gaps in Parents' Discipline Strategies from 1988 to 2011," *Pediatrics* 138 (2016): e20160720, https://app.dimensions.ai/details/publication/pub.1067831611, and U.S. Department of Education, "Educators Gather at the White House to Rethink School Discipline," July 22, 2015, https://www.ed.gov/news/press-releases/educators-gather-white-house-rethink-school-discipline.

117 **self-esteem movement:** R. F. Baumeister, J. D. Campbell, J. I. Krueger, and K. D. Vohs, "Does High Self-Esteem Cause Better Performance, Interpersonal Success, Happiness, or Healthier Lifestyles?," *Psychological Science in the Public Interest* 4 (2003): 1–44.

118 **guide for teachers:** S. Zemelman, H. S. Daniels, and A. Hyde, *Best Practice,* 4th ed., *Bringing Standards to Life in America's Classrooms* (Portsmouth, NH: Heinemann, 2012).

118 **Herbert Marsh tested:** H. W. Marsh et al., "Dimensional Comparison Theory: An Extension of the Internal/External Frame of Reference Effect on Academic Self-Concept Formation," *Contemporary Educational Psychology* 39 (2014): 326–41, and H. W. Marsh et al., "Long-Term

Positive Effects of Repeating a Year in School: Six-Year Longitudinal Study of Self-Beliefs, Anxiety, Social Relations, School Grades, and Test Scores," *Journal of Educational Psychology* 109 (2017): 425–38.

118 **students in Florida:** J. P. Greene and M. A. Winters, "Revisiting Grade Retention: An Evaluation of Florida's Test-Based Promotion Policy," *Education Finance and Policy* 2 (2007): 319–40, https://www.mitpressjournals.org/doi/pdf/10.1162/edfp.2007.2.4.319.

118 **schools are reluctant:** K. Spencer, "A New Kind of Classroom: No Grades, No Failing, No Hurry," *New York Times,* August 11, 2017, https://www.nytimes.com/2017/08/11/nyregion/mastery-based-learning-no-grades.html, and D. Rado, "Parents Push Back Against School Report Cards with No Letter Grades," *Chicago Tribune,* June 6, 2016, http://www.chicagotribune.com/news/local/breaking/ct-middle-school-grades-met-20160601-story.html.

119 **High-school grades:** L. Vries, "High School Grades Hit by Inflation," CBS News, January 27, 2003, https://www.cbsnews.com/news/high-school-grades-hit-by-inflation/, and "Average High School GPAs Increased Since 1990," *U.S. News & World Report,* April 19, 2011, https://www.usnews.com/opinion/articles/2011/04/19/average-high-school-gpas-increased-since-1990.

119 **with class rank:** M. Balingit, "High Schools Are Doing Away with Class Rank. What Does That Mean for College Admissions?," *Washington Post,* July 13, 2015, https://www.washingtonpost.com/news/grade-point/wp/2015/07/13/high-schools-are-doing-away-with-class-rank-what-does-that-mean-for-college-admissions/?utm_term=.1383695e1064.

119 **are floundering academically:** N. M. Fortin, P. Oreopoulos, and S. Phipps, "Leaving Boys Behind: Gender Disparities in High Academic Achievement," *Journal of Human Resources* 50 (2015): 549–79; and Pew Research Center, "Women Outpace Men in College Enrollment," Fact Tank, March 6, 2014, http://www.pewresearch.org/fact-tank/2014/03/06/womens-college-enrollment-gains-leave-men-behind/; J. Guo, "Poor Boys Are Falling Behind Poor Girls, and It's Deeply Troubling," *Washington Post,* November 23, 2015, https://www.washingtonpost.com/news/wonk/wp/2015/11/23/why-girls-do-so-much-better-than-boys-in-school/?utm_term=.b6aafa26dd38.

119 **develop self-control:** A. L. Duckworth and M. E. P. Seligman, "Self-Discipline Gives Girls the Edge: Gender in Self-Discipline, Grades, and Achievement Test Scores," *Journal of Educational Psychology* 98 (2006): 198–208, https://dx.doi.org/10.1037/0022-0663.98.1.198.

119 **boys in homes:** R. F. Baumeister and J. Tierney, *Willpower: Rediscovering the Greatest Human Strength* (New York: Penguin Press, 2011), 207–10.

119 **consequences for boys:** D. Autor, D. Figlio, K. Karbownik, J. Roth, and M. Wasserman, "Family Disadvantage and the Gender Gap in Behavioral and Educational Outcomes," National Bureau of Economic Research, NBER Working Paper No. 22267 (2017), and M. Bertrand and J. Pan, "The Trouble with Boys: Social Influences and the Gender Gap in Disruptive Behavior," National Bureau of Economic Research, NBER Working Paper No. 17541 (2011).

120 **a 2000 book:** S. C. Carter, *No Excuses: Lessons from 21 High-Performing, High-Poverty Schools* (Washington, DC: Heritage Foundation, 2000).

120 **a subsequent book:** A. Thernstrom and S. Thernstrom, *No Excuses: Closing the Racial Gap in Learning* (New York: Simon & Schuster, 2003).

120 **At Success Academy:** Sources include interviews with Eva Moskowitz and Ann Powell of Success Academy as well as E. Moskowitz, *The Education of Eva Moskowitz: A Memoir* (New York: HarperCollins, 2017); R. Mead, "Success Academy's Radical Educational Experiment," *New Yorker,* December 11, 2017, https://www.newyorker.com/magazine/2017/12/11/success-academys-radical-educational-experiment; K. Taylor, "At Success Academy Charter Schools, High Scores and Polarizing Tactics," *New York Times,* April 6, 2015, https://www.nytimes.com/2015/04/07/nyregion/at-success-academy-charter-schools-polarizing-methods-and-superior-results.html; and Success Academy Charter Schools, https://www.successacademies.org/.

122 **state proficiency exams:** New York State Education Department, "Measuring Student Proficiency in Grades 3–8 English Language Arts and Mathematics," September 26, 2018, http://www.nysed.gov/common/nysed/files/programs/information-reporting-services/2018-3-8-test-results-presentation.pdf.

123 **2017 meta-analysis:** A. Cheng, C. Hitt, B. Kisida, and J. N. Mills, "'No Excuses' Charter Schools: A Meta-analysis of the Experimental Evidence on Student Achievement," *Journal of School Choice* 11 (2017): 209–38. See also J. D. Angrist, P. A. Pathak, and C. R. Walters, "Explaining Charter School Effectiveness," *American Economic Journal: Applied Economics* 5 (2013): 1–27.

124 **retains a higher percentage:** J. Roy, "Staying or Going? Comparing Student Attrition Rates at Charter Schools with Nearby Traditional Public Schools," New York City Independent Budget Office Schools Brief, January 2014, http://www.ibo.nyc.ny.us/iboreports/2014attritioncharterpublic.html?mod=article_inline; B. Fertig and J. Ye, "NYC Charters Retain Students Better Than Traditional Schools," WNYC, March 15, 2016, https://www.wnyc.org/story/nyc-charter-school-attrition-rates/, and E. S. Moskowitz, "The Myth of Charter-School 'Cherry Picking,'" *Wall Street Journal,* February 8, 2015, https://www.wsj.com/articles/eva-s-moskowitz-the-myth-of-charter-school-cherry-picking-1423438046.

124 **torrent of criticism:** K. Taylor, "At Success Academy School, a Stumble in Math and a Teacher's Anger on Video," *New York Times,* February 12, 2016, https://www.nytimes.com/2016/02/13/nyregion/success-academy-teacher-rips-up-student-paper.html?_r=0; E. Green, "Beyond the Viral Video: Inside Educators' Emotional Debate About 'No Excuses' Discipline," Chalkbeat New York, March 8, 2016, https://ny.chalkbeat.org/posts/ny/2016/03/08/beyond-the-viral-video-inside-educators-emotional-debate-about-no-excuses-discipline/; and Moskowitz, *Education of Eva Moskowitz,* 320–22.

124 **only twenty-seven hours:** P. S. Babcock and M. Marks, "The Falling Time Cost of College: Evidence from Half a Century of Time Use Data," *Review of Economics and Statistics* 93 (2011): 468–78, http://faculty.ucr.edu/~mmarks/Papers/babcock2010falling.pdf.

124 **grade is an A:** S. Rojstaczer, "Grade Inflation at American Colleges and Universities," GradeInflation.com, March 29, 2016, http://www.gradeinflation.com; S. Rojstaczer and C. Healy, "Where A Is Ordinary: The Evolution of American College and University Grading, 1940–2009," *Teachers College Record* 114 (2012): 1–23; and C. Rampell, "The Rise of the 'Gentleman's A' and the GPA Arms Race," *Washington Post,*

March 28, 2016, https://www.washingtonpost.com/opinions/the-rise-of-the-gentlemans-a-and-the-gpa-arms-race/2016/03/28/05c9e966-f522-11e5-9804-537defcc3cf6_story.html?utm_term=.73a99f546f4b.

124 **study in 2011:** R. Arum and J. Roksa, *Academically Adrift: Limited Learning on College Campuses* (Chicago: University of Chicago Press, 2011), Kindle.

125 **follow-up study:** R. Arum and J. Roksa, *Aspiring Adults Adrift: Tentative Transitions of College Graduates* (Chicago: University of Chicago Press, 2014), and K. Carey, "The Economic Price of Colleges' Failures," *New York Times,* September 2, 2014, https://www.nytimes.com/2014/09/03/upshot/the-economic-price-of-colleges-failures.html?rref=upshot&abt=0002&abg=0&_r=0.

125 **efforts have faltered:** J. Saffron, "Reforms Aimed at Fighting Grade Inflation Are Falling Short," James G. Martin Center, May 4, 2015, https://www.jamesgmartin.center/2015/05/reforms-aimed-at-fighting-grade-inflation-are-falling-short/.

125 **"truth in grading":** T. K. Lindsay, "If A Is Average, Say So—the Dawn of Honest Transcripts," Minding the Campus, December 18, 2013, https://www.mindingthecampus.org/2013/12/18/if_a_is_average_say_so-the_da/.

126 **legislators in Texas:** T. Lindsay, "Texas Legislature Looks to Lift College Grading Standards," *Forbes,* February 28, 2015, https://www.forbes.com/sites/tomlindsay/2015/02/28/the-texas-legislature-looks-to-lift-college-grading-standards/#52e0bd41a037.

126 **like Harvey Mansfield:** T. Andersen, N. Jacques, and T. Feathers, "Harvard Professor Says Grade Inflation Rampant," *Boston Globe,* December 4, 2013, https://www.bostonglobe.com/metro/2013/12/03/harvard-professor-raises-concerns-about-grade-inflation/McZHfRZ2RxpoP5XvwgedlN/story.html, and M. Q. Clarida and N. P. Fandos, "Substantiating Fears of Grade Inflation, Dean Says Median Grade at Harvard College is A–, Most Common Grade Is A," *Harvard Crimson,* May 26, 2017, https://www.thecrimson.com/article/2013/12/3/grade-inflation-mode-a/.

126 **city near Chicago:** R. G. Fryer, S. D. Levitt, J. List, and S. Sadoff, "Enhancing the Efficacy of Teacher Incentives Through Loss Aversion: A

Field Experiment," National Bureau of Economic Research, NBER Working Paper No. 18237 (2012).

127 **factory in Nanjing:** T. Hossain and J. A. List, "The Behavioralist Visits the Factory: Increasing Productivity Using Simple Framing Manipulations," *Management Science* 58 (2012): 2151–67.

128 **Potato-Chip Awakening:** Drawn from an interview with Dick Grote and from his book *Discipline Without Punishment: The Proven Strategy That Turns Problem Employees into Superior Performers* (New York: AMACOM, 2006).

132 **psychologists have repeatedly demonstrated:** R. F. Baumeister, A. M. Stillwell, and T. F. Heatheron, "Guilt: An Interpersonal Approach," *Psychological Bulletin* 115 (1994): 243–67; J. P. Tangney and R. L. Dearing, *Shame and Guilt* (New York: Guilford Press, 2002); F. J. Flynn and R. L. Schaumberg, "When Feeling Bad Leads to Feeling Good: Guilt-Proneness and Affective Organizational Commitment," *Journal of Applied Psychology* (2012): 124–33.

CHAPTER 6: Business 101

136 **Byington began discussing:** Interviews with Eliza Byington and Will Felps.

136 **linguist Geoff Nunberg:** G. Nunberg, "Bad Apple Proverbs: There's One in Every Bunch," *Fresh Air*, NPR, May 5, 2011, https://www.npr.org/2011/05/09/136017612/bad-apple-proverbs-theres-one-in-every-bunch.

137 **Benjamin Franklin warned:** B. Franklin, *Poor Richard's Almanack* (Philadelphia: New Printing Office, 1736), Kindle.

137 **one "roten appul":** G. Chaucer, "The Cook's Tale," in *The Canterbury Tales* (Ware, UK: Wordsworth Editions, 2002), 151.

137 ***Social support* became:** J. Holt-Lunstad, T. B. Smith, and J. B. Layton, "Social Relationships and Mortality Risk: A Meta-analytic Review," *PLoS Med* 7 (2010): e1000316, https://doi.org/10.1371/journal.pmed.1000316; J. Lynch, *The Broken Heart: The Medical Consequences of Loneliness* (New York: Basic Books, 1977); and J. T. Cacioppo and

W. Patrick, *Loneliness: Human Nature and the Need for Social Connection* (New York: W. W. Norton, 2008).

138 **study of widows:** K. S. Rook, "The Negative Side of Social Interaction: Impact on Psychological Well-Being," *Journal of Personality and Social Psychology* 46 (1984): 1097–108.

138 **Other studies of elderly:** M. A. Okun, J. F. Melichar, and M. D. Hill, "Negative Daily Events, Positive and Negative Social Ties, and Psychological Distress Among Older Adults," *Gerontologist* 30 (1990): 193–99; J. F. Finch, M. A. Okun, M. Barrera, A. J. Zautra, and J. W. Reich, "Positive and Negative Social Ties Among Older Adults: Measurement Models and the Prediction of Psychological Distress and Well-Being," *American Journal of Community Psychology* 17 (1989): 585–605, https://doi.org/10.1007/BF00922637; and J. T. Newsom, K. S. Rook, M. Nishishiba, D. H. Sorkin, and T. L. Mahan, "Understanding the Relative Importance of Positive and Negative Social Exchanges: Examining Specific Domains and Appraisals," *Journals of Gerontology Series B: Psychological Sciences and Social Sciences* 60 (2005): 304–12, https://www.ncbi.nlm.nih.gov/pmc/articles/PMC3833824/.

138 **of unemployed people:** A. D. Vinokur and M. van Ryn, "Social Support and Undermining in Close Relationships: Their Independent Effects on the Mental Health of Unemployed Persons," *Journal of Personality and Social Psychology* 65 (1993): 350–59.

139 **One study, titled:** M. K. Duffy, D. Ganster, and M. Pagon, "Social Undermining in the Workplace," *Academy of Management Journal* 45 (2002): 331–51, https://doi.org/10.2307/3069350.

139 **chain in Australia:** P. D. Dunlop and K. Lee, "Workplace Deviance, Organizational Citizenship Behavior, and Business Unit Performance: The Bad Apples Do Spoil the Whole Barrel," *Journal of Organizational Behavior* 25 (2004): 67–80, https://dx.doi.org/10.1002/job.243.

140 **with job performance:** G. M. Hurtz and J. J. Donovan, "Personality and Job Performance: The Big Five Revisited," *Journal of Applied Psychology* 85 (2000): 869–79.

141 **four-person teams:** L. M. Camacho and P. B. Paulus, "The Role of Social Anxiousness in Group Brainstorming," *Journal of Personality and Social Psychology* 68 (1995): 1071–80.

141 **several manufacturing companies:** M. R. Barrick, G. L. Stewart, M. J. Neubert, and M. K. Mount, "Relating Member Ability and Personality to Work-Team Processes and Team Effectiveness," *Journal of Applied Psychology* 83 (1998): 377–91.

142 **overview of the problem:** W. Felps, T. Mitchell, and E. Byington, "How, When, and Why Bad Apples Spoil the Barrel: Negative Group Members and Dysfunctional Groups," *Research in Organizational Behavior* 27 (2006): 175–222, and S. Heen, "Bad Apple Behavior and Its Impact on Team Results," White Paper No. 2, Triad Institute, https://triadconsultinggroup.com/sites/default/files/Triad_Consulting_Whitepaper_2_Case_Study_%20114_Bad_Apples.pdf.

143 **is continually glum:** S. Kaplan, J. C. Bradley, J. N. Luchman, and D. Haynes, "On the Role of Positive and Negative Affectivity in Job Performance: A Meta-analytic Investigation," *Journal of Applied Psychology* 94 (2009): 162–76.

143 **feeling bad afterward:** M. J. Howes, J. E. Hokanson, and D. A. Loewenstein, "Induction of Depressive Affect After Prolonged Exposure to a Mildly Depressed Individual," *Journal of Personality and Social Psychology* 49 (1985): 1110–13.

143 **firm in the Midwest:** A. G. Miner, T. M. Glomb, and C. Hulin, "Experience Sampling Mood and Its Correlates at Work," *Journal of Occupational and Organizational Psychology* 78 (2005): 171–93.

144 **by Warren Jones:** Personal communication.

145 **paper by Baumeister:** R. F. Baumeister and M. R. Leary, "The Need to Belong: Desire for Interpersonal Attachments as a Fundamental Human Motivation," *Psychological Bulletin* 117 (1995): 497–529, https://dx.doi.org/10.1037/0033-2909.117.3.497.

145 **collaborator in Jean Twenge:** R. F. Baumeister, L. E. Brewer, D. M. Tice, and J. M. Twenge, "Thwarting the Need to Belong: Understanding the Interpersonal and Inner Effects of Social Exclusion," *Social and Personality Psychology Compass* 1 (2007): 506–20.

145 **social death sentence:** J. M. Twenge, K. R. Catanese, and R. F. Baumeister, "Social Exclusion Causes Self-Defeating Behavior," *Journal of Personality and Social Psychology* 83 (2002): 606–15, https://doi.org/10.1037/0022-3514.83.3.606.

146 **several hundred studies:** C. N. DeWall and B. J. Bushman, "Social Acceptance and Rejection: The Sweet and the Bitter," *Current Directions in Psychological Science* 20 (2011): 256–60.

146 **"North Pond Hermit":** M. Finkel, *The Stranger in the Woods* (New York: Knopf, 2017), 62.

147 **self-control suffers:** R. F. Baumeister, C. N. DeWall, N. J. Ciarocco, and J. M. Twenge, "Social Exclusion Impairs Self-Regulation," *Journal of Personality and Social Psychology* 88 (2005): 589–604, https://doi.org/10.1037/0022-3514.88.4.589.

147 **tests of intelligence:** R. F. Baumeister, J. M. Twenge, and C. K. Nuss, "Effects of Social Exclusion on Cognitive Processes: Anticipated Aloneness Reduces Intelligent Thought," *Journal of Personality and Social Psychology* 83 (2002): 817–27, https://doi.org/10.1037/0022-3514.83.4.817.

147 **was relatively muted:** J. M. Twenge, K. R. Catanese, and R. F. Baumeister, "Social Exclusion and the Deconstructed State: Time Perception, Meaninglessness, Lethargy, Lack of Emotion, and Self-Awareness," *Journal of Personality and Social Psychology* 85 (2003): 409–23, https://doi.org/10.1037/0022-3514.85.3.409, and G. C. Blackhart, B. C. Nelson, M. L. Knowles, and R. F. Baumeister, "Rejection Elicits Emotional Reactions but Neither Causes Immediate Distress nor Lowers Self-Esteem: A Meta-analytic Review of 192 Studies on Social Exclusion," *Personality and Social Psychology Review* 13 (2009): 269–309.

147 **their heart rate:** B. Gunther Moor, E. A. Crone, and M. W. van der Molen, "The Heartbrake of Social Rejection: Heart Rate Deceleration in Response to Unexpected Peer Rejection," *Psychological Science* 21 (2010): 1326–33.

147 **physical senses dulled:** C. N. DeWall and R. F. Baumeister, "Alone but Feeling No Pain: Effects of Social Exclusion on Physical Pain Tolerance and Pain Threshold, Affective Forecasting, and Interpersonal Empathy," *Journal of Personality and Social Psychology* 91 (2006): 1–15, https://doi.org/10.1037/0022-3514.91.1.1.

147 **game called Cyberball:** N. I. Eisenberger, M. D. Lieberman, and K. D. Williams, "Does Rejection Hurt? An fMRI Study of Social Exclusion," *Science* 302 (2003): 290–92. See also N. I. Eisenberger, J. M. Jarcho, M. Lieberman, and B. D Naliboff, "An Experimental Study of

Shared Sensitivity to Physical Pain and Social Rejection," *Pain* 126 (2006): 132–38, https://doi.org/10.1016/j.pain.2006.06.024.

148 **a romantic partner:** H. E. Fisher et al., "Reward, Addiction, and Emotion Regulation Systems Associated with Rejection in Love," *Journal of Neurophysiology* 104 (2010): 51–60, and E. Kross, M. G. Berman, W. Mischel, E. E. Smith, and T. D. Wager, "Social Rejection Shares Somatosensory Representations with Physical Pain," *Proceedings of the National Academy of Sciences* 108 (2011): 6270–75, https://doi.org/10.1073/pnas.1102693108.

148 **giving people Tylenol:** C. N. DeWall et al., "Acetaminophen Reduces Social Pain: Behavioral and Neural Evidence," *Psychological Science* 21 (2010): 931–37.

148 **use of marijuana:** T. Deckman, C. N. DeWall, R. Gilman, B. Way, and S. Richman, "Can Marijuana Reduce Social Pain?," *Social Psychological and Personality Science* 5 (2014): 131–39.

148 **significantly less helpful:** J. M. Twenge, R. F. Baumeister, C. N. DeWall, N. J. Ciarocco, and J. M. Bartels, "Social Exclusion Decreases Prosocial Behavior," *Journal of Personality and Social Psychology* 92 (2007): 56–66, https://doi.org/10.1037/0022-3514.92.1.56.

149 **"blood-covered glasses":** C. N. DeWall, J. M. Twenge, S. A. Gitter, and R. F. Baumeister, "It's the Thought That Counts: The Role of Hostile Cognition in Shaping Aggressive Responses to Social Exclusion," *Journal of Personality and Social Psychology* 96 (2009): 45–59.

150 **shootings at schools:** M. R. Leary, R. M. Kowalski, L. Smith, and S. Phillips, "Teasing, Rejection, and Violence: Case Studies of the School Shootings," *Aggressive Behavior* 29 (2003): 202–14.

150 **unite to retaliate:** I. van Beest, A. R. Carter-Sowell, E. van Dijk, and K. D. Williams, "Groups Being Ostracized by Groups: Is the Pain Shared, Is Recovery Quicker, and Are Groups More Likely to Be Aggressive?," *Group Dynamics: Theory, Research, and Practice* 16 (2012): 241–54, https://dx.doi.org/10.1037/a0030104.

151 **subsequent popular book:** R. I. Sutton, *The No Asshole Rule* (New York: Hachette, 2007), Kindle.

152 **Big Five personality:** J. Hogan, P. Barrett, and R. Hogan, "Personality Measurement, Faking, and Employment Selection," *Journal of Applied*

Psychology 92 (2007): 1270–85; J. Hogan and B. Holland, "Using Theory to Evaluate Personality and Job-Performance Relations: A Socioanalytic Perspective," *Journal of Applied Psychology* 88 (2003): 100–112; and M. R. Barrick and M. K. Mount, "The Big Five Personality Dimensions and Job Performance: A Meta-analysis," *Personnel Psychology* 44 (1991): 1–26.

153 **the "triangle hypothesis":** H. H. Kelley and A. J. Stahelski, "Social Interaction Basis of Cooperators' and Competitors' Beliefs About Others," *Journal of Personality and Social Psychology* 16 (1970): 66–91.

154 **young Steve Jobs:** W. Isaacson, *Steve Jobs* (New York: Simon & Schuster, 2011), 43.

155 **At Men's Wearhouse:** Sutton, *No Asshole Rule,* chap. 3, Kindle.

CHAPTER 7: Online Perils

159 **The Sunshine Hotel:** Material about the Casablanca Hotel is drawn from interviews with Adele Gutman, vice president of the Library Hotel Collection, and John Taboada, general manager of the Casablanca Hotel, as well as https://casablancahotel.com/ and reviews and ratings at TripAdvisor, https://www.tripadvisor.com/Hotel_Review-g60763-d113317-Reviews-Casablanca_Hotel_by_Library_Hotel_Collection-New_York_City_New_York.html.

161 **"I get a tearful":** Interview with Adryenn Ashley.

162 **tested its impact:** M. Lee and S. Youn, "Electronic Word of Mouth (eWOM): How eWOM Platforms Influence Consumer Product Judgement," *International Journal of Advertising* 28 (2009): 473–99, https://vdocuments.site/electronic-word-of-mouth-ewom-how-ewom-platforms-influence-consumer-product-judgement.html.

163 **planning a vacation:** A. Papathanassis and F. Knolle, "Exploring the Adoption and Processing of Online Holiday Reviews: A Grounded Theory Approach," *Tourism Management* 32 (2011): 215–24.

163 **people are swayed:** J. Lee, D. Park, and I. Han, "The Effect of Negative Online Consumer Reviews on Product Attitude: An Information

Processing Review," *Electronic Commerce Research and Applications* 7 (2008): 341–52.

163 **trends at Amazon:** J. A. Chevalier and D. Mayzlin, "The Effect of Word of Mouth on Sales: Online Book Reviews," *Journal of Marketing Research* 43 (2006): 345–54.

163 **old-fashioned word of mouth:** J. Goodman and S. Newman, "Understand Customer Behavior and Complaints," *Quality Progress* 36 (2003): 51–55; E. W. Anderson, "Customer Satisfaction and Word of Mouth," *Journal of Service Research* 1 (1998): 5–17, https://doi.org/10.1177/109467059800100102; R. East, K. Hammond, and M. Wright, "The Relative Incidence of Positive and Negative Word of Mouth: A Multi-category Study," *International Journal of Research in Marketing* 24 (2007): 175–84; and T. O. Jones and W. E. Sasser, "Why Satisfied Customers Defect," *Harvard Business Review* (November—December 1995), https://hbr.org/1995/11/why-satisfied-customers-defect.

163 **1980s at Xerox:** J. L. Heskett, W. E. Sasser, and L. A. Schlesinger, *The Service Profit Chain: How Leading Companies Link Profit and Growth to Loyalty, Satisfaction, and Value* (New York: Free Press, 1997), Kindle.

164 **Harvard Business School:** Jones and Sasser, "Why Satisfied Customers Defect"; and Heskett, Sasser, and Schlesinger, *Service Profit Chain.*

164 **musician Dave Carroll:** Gulliver, "Did Dave Carroll Lose United Airlines $180m?," *Economist,* July 24, 2009, https://www.economist.com/gulliver/2009/07/24/did-dave-carroll-lose-united-airlines-180m.

164 **wrote a song:** D. Carroll, "United Breaks Guitars," YouTube, https://www.youtube.com/watch?v=5YGc4zOqozo.

165 **When McDonald's wanted:** K. Hill, "#McDStories: When a Hashtag Becomes a Bashtag," *Forbes,* January 24, 2012, https://www.forbes.com/sites/kashmirhill/2012/01/24/mcdstories-when-a-hashtag-becomes-a-bashtag/#5eac3887ed25.

165 **average review rating:** ReviewTrackers, "2018 ReviewTrackers Online Reviews Survey: Statistics and Trends," https://www.reviewtrackers.com/online-reviews-survey/, and G. A. Fowler, "When 4.3 Stars Is Average: The Internet's Grade-Inflation Problem," *Wall Street Journal,* April 5, 2017, https://www.wsj.com/articles/when-4-3-stars-is-average-the-internets-grade-inflation-problem-1491414200.

165 **"public self-enhancement":** A. C. Wojnicki and D. Godes, "Signaling Success: Word of Mouth as Self-Enhancement," *Customer Needs and Solutions* 4 (2017): 68–82.

166 **"On Braggarts and Gossips":** M. De Angelis, A. Bonezzi, A. M. Peluso, D. D. Rucker, and M. Costabile, "On Braggarts and Gossips: A Self-Enhancement Account of Word-of-Mouth Generation and Transmission," *Journal of Marketing Research* 49 (2012): 551–63, https://doi.org/10.1509/jmr.11.0136.

166 **clay-animation film:** A. E. Schlosser, "Posting Versus Lurking: Communicating in a Multiple Audience Context," *Journal of Consumer Research* 32 (2005): 260–65, https://doi.org/10.1086/432235.

167 **ratings tend to drop:** X. Li and L. M. Hitt, "Self-Selection and Information Role of Online Product Reviews," *Information Systems Research* 19 (2008): 456–74, https://dx.doi.org/10.1287/isre.1070.0154; W. Moe and M. Trusov, "The Value of Social Dynamics in Online Product Ratings Forums," *Journal of Marketing Research* 48 (2011): 444–56, https://doi.org/10.1509/jmkr.48.3.444; and D. Godes and J. Silva, "Sequential and Temporal Dynamics of Online Opinion," *Marketing Science* 31 (2012): 448–73, https://doi.org/10.1287/mksc.1110.0653.

167 **minority of activists:** W. Moe and D. Schweidel, "Online Product Opinions: Incidence, Evaluation, and Evolution," *Marketing Science* 31 (2012): 372–86.

167 **fifth of Americans:** YouGov Omnibus Survey, January 2014, https://today.yougov.com/topics/lifestyle/articles-reports/2014/01/22/21-americans-have-reviewed-products-and-services-t.

167 **study comparing reviews:** E. T. Anderson and D. I. Simester, "Reviews Without a Purchase: Low Ratings, Loyal Customers, and Deception," *Journal of Marketing Research* 51 (2014): 249–69, https://doi.org/10.1509/jmr.13.0209.

167 **a few signs:** Anderson and Simester, "Reviews Without a Purchase," and M. Ott, Y. Choi, C. Cardie, and J. Hancock, "Finding Deceptive Opinion Spam by Any Stretch of the Imagination," *Proceedings of the 49th Annual Meeting of the Association for Computational Linguistics* (Stroudsburg, PA: Association for Computational Linguistics, 2011): 309–19.

167 **Amazon's Canadian website:** A. Harmon, "Amazon Glitch Unmasks War of Reviewers," *New York Times,* February 14, 2004, https://www.nytimes.com/2004/02/14/us/amazon-glitch-unmasks-war-of-reviewers.html.

168 **ad on Craigslist:** M. Willett, "A Craigslist Ad Is Offering $25 for Fake Yelp Reviews of NYC Restaurants," *Business Insider,* May 14, 2013.

168 **for $5 apiece:** K. Knibbs, "If You Can't Spare $25 for a Fake Yelp Review, There's Plenty Available for $5," *Digital Trends,* May 14, 2013, https://www.digitaltrends.com/social-media/fake-yelp-reviews/#!K80fU.

168 **a British consultant:** Interview with Chris Emmins, director and co-founder of KwikChex, https://kwikchex.com/, and O. Smith, "TripAdvisor Under Fire over Fraud Detection," *Telegraph,* March 22, 2012, https://www.telegraph.co.uk/travel/news/TripAdvisor-under-fire-over-fraud-detection/.

168 **by Dina Mayzlin:** D. Mayzlin, Y. Dover, and J. Chevalier, "Promotional Reviews: An Empirical Investigation of Online Review Manipulation," *American Economic Review* 104 (2014): 2421–55.

169 **10 percent less:** C. Anderson, "The Impact of Social Media on Lodging Performance," *Cornell Hospitality Report* 12 (2012): 6–11; H. Öğüt and B. Taş, "The Influence of Internet Customer Reviews on Online Sales and Prices in the Hotel Industry," *Service Industries Journal* 32 (2012): 197–214; and E. N. Torres, D. Singh, and A. Robertson-Ring, "Consumer Reviews and the Creation of Booking Transaction Value: Lessons from the Hotel Industry," *International Journal of Hospitality Management* 50 (2015): 77–83.

169 **owners of Botto Bistro:** Botto Bistro Facebook page, https://www.facebook.com/bottobistro/, and Botto Bistro Yelp page, https://www.yelp.com/biz/botto-italian-bistro-richmond-9.

170 **business-school talk:** A. Gutman, "Library Hotel Collection: 5 Secrets to Brand Reputation Building," *ReviewPro Case Studies,* January 21, 2019, https://www.reviewpro.com/blog/category/case-studies/, and A. Gutman, "Best Practices in Reputation Management: Whose Job Is It Anyway?," *Hotel Business Review,* February 17, 2013, https://www.hotelexecutive.com/business_review/3353/best-practices-in-reputation-management-whose-job-is-it-anyway.

172 **administering job interviews:** B. I. Bolster and B. M. Springbett, "The Reaction of Interviewers to Favorable and Unfavorable Information," *Journal of Applied Psychology* 45 (1961): 97–103.

172 **in ice water:** B. L. Fredrickson and D. Kahneman, "Duration Neglect in Retrospective Evaluations of Affective Episodes," *Journal of Personality and Social Psychology* 65 (1993): 45–55.

172 **of patients' reactions:** D. A. Redelmeier and D. Kahneman, "Patients' Memories of Painful Medical Treatments: Real-Time and Retrospective Evaluations of Two Minimally Invasive Procedures," *Pain* 66 (1996): 3–8, https://doi.org/10.1016/0304-3959(96)02994-6, and D. A. Redelmeier, J. Katz, and D. Kahneman, "Memories of Colonoscopy: A Randomized Trial," *Pain* 104 (2003): 187–94, https://doi.org/10.1016/S0304-3959(03)00003-4.

172 **researchers from Dartmouth:** A. M. Do, A. V. Rupert, and G. Wolford, "Evaluations of Pleasurable Experiences: The Peak-End Rule," *Psychonomic Bulletin & Review* 15 (2008): 96–98.

173 **researchers in South Korea:** H. Lee and R. S. Dalal, "The Effects of Performance Extremities on Ratings of Dynamic Performance," *Human Performance* 24 (2011): 99–118.

176 **responses on TripAdvisor:** C. Anderson and S. Han, "Hotel Performance Impact of Socially Engaging with Consumers," *Cornell Hospitality Report* 16 (2016): 3–9.

CHAPTER 8: The Pollyanna Principle

179 **the 1913 novel:** E. H. Porter, *Pollyanna* (Boston: Louis Coues Page, 1913).

179 **an instant bestseller:** R. Graham, "How We All Became Pollyannas (and Why We Should Be Glad About It)," *Atlantic,* February 26, 2013, https://www.theatlantic.com/entertainment/archive/2013/02/how-we-all-became-pollyannas-and-why-we-should-be-glad-about-it/273323/.

180 **Marion later noted:** F. Marion, *Off with Their Heads! A Serio-Comic Tale of Hollywood* (New York: Macmillan, 1972), 67.

181 **D. W. Griffith:** C. Keil, ed., *A Companion to D. W. Griffith* (Hoboken, NJ: Wiley, 2018), 61.

181 **Studies in Chicago:** R. W. Schrauf and J. Sanchez, "The Preponderance of Negative Emotion Words in the Emotion Lexicon: A Cross-Generational and Cross-Linguistic Study," *Journal of Multilingual and Multicultural Development* 25 (2004): 266–84, https://doi.org/10.1080/01434630408666532.

181 **half-dozen European:** S. Van Goozen and N. H. Frijda, "Emotion Words Used in Six European Countries," *European Journal of Social Psychology* 23 (1993): 89–95.

181 **pored through dictionaries:** J. R. Averill, "On the Paucity of Positive Emotions," in *Advances in the Study of Communication and Affect*, vol. 6, ed. K. Blankstein, P. Pliner, and J. Polivy (New York: Plenum, 1980), 745, and J. A. Russell, "Culture and the Categorization of Emotions," *Psychological Bulletin* 110 (1991): 426–50, https://dx.doi.org/10.1037/0033-2909.110.3.426.

181 **count the words:** T. Bontranger, "The Development of Word Frequency Lists Prior to the 1944 Thorndike-Lorge List," *Reading Psychology* 12 (1991): 91–116; R. B. Zajonc, "Attitudinal Effects of Mere Exposure," *Journal of Personality and Social Psychology* 9 (1968): 1–27; and E. L. Thorndike and I. Lorge, *The Teacher's Wordbook of 30,000 Words* (New York: Teachers College, Columbia University, 1944).

182 **paper in 1969:** J. Boucher and C. E. Osgood, "The Pollyanna Hypothesis," *Journal of Verbal Learning and Verbal Behavior* 8 (1969): 1–8.

182 **Pollyannaism reigned in:** I. M. Kloumann, C. M. Danforth, K. D. Harris, C. A. Bliss, and P. S. Dodds, "Positivity of the English Language," *PLoS ONE* 7 (2012): e29484, https://doi.org/10.1371/journal.pone.0029484. See also a study that used Google's web crawler to detect a positivity bias in the frequency of emotionally laden words on English, German, and Spanish websites: D. Garcia, A. Garas, and F. Schweitzer, "Positive Words Carry Less Information Than Negative Words," *EPJ Data Science* 1 (2012): 3, https://arxiv.org/abs/1110.4123.

182 **called a hedonometer:** https://hedonometer.org/index.html.

182 **in ten languages:** P. S. Dodds et al., "Human Language Reveals a Universal Positivity Bias," *Proceedings of the National Academy of Sciences*

112 (2015): 2389–94, https://doi.org/10.1073/pnas.1411678112, and J. Tierney, "According to the Words, the News Is Actually Good," *New York Times,* February 23, 2015, https://www.nytimes.com/2015/02/24/science/why-we-all-sound-like-pollyannas.html.

183 **If today is bad:** K. Sheldon, R. Ryan, and H. Reis, "What Makes for a Good Day? Competence and Autonomy in the Day and in the Person," *Personality and Social Psychology Bulletin* 22 (1996): 1270–79.

183 **tell longer stories:** A. Abele, "Thinking About Thinking: Causal, Evaluative and Finalistic Cognitions About Social Situations," *European Journal of Social Psychology* 15 (1985): 315–32.

183 **By digitally altering facial:** J. Golle, F. W. Mast, and J. S. Lobmaier, "Something to Smile About: The Interrelationship Between Attractiveness and Emotional Expression," *Cognition and Emotion* 28 (2014): 298–310.

184 **gained more followers:** C. J. Hutto, S. Yardi, and E. Gilbert, "A Longitudinal Study of Follow Predictors on Twitter," in *Proceedings of the SIGCHI Conference on Human Factors in Computing Systems (CHI '13)* (New York: ACM, 2013), 821–30, https://doi.org/10.1145/2470654.2470771.

184 **adopt upbeat themes:** E. Ferrara and Z. Yang, "Measuring Emotional Contagion in Social Media," *PLoS ONE* 10 (2015): e0142390, https://doi.org/10.1371/journal.pone.0142390.

184 **spread more widely:** E. Ferrara and Z. Yang, "Quantifying the Effect of Sentiment on Information Diffusion in Social Media," *PeerJ Computer Science* 1 (2015): e26, https://doi.org/10.7717/peerj-cs.26.

184 **British Royal Society:** Royal Society for Public Health, "#StatusOfMind: Social Media and Young People's Mental Health and Wellbeing," London, 2017, https://www.rsph.org.uk/uploads/assets/uploaded/62be270a-a55f-4719-ad668c2ec7a74c2a.pdf.

184 **time on YouTube:** DMR, "160 Amazing YouTube Statistics and Facts," December 2018, https://expandedramblings.com/index.php/youtube-statistics/.

185 **comic-book scare:** F. Wertham, *Seduction of the Innocent* (New York: Rinehart, 1954); F. Wertham, "Comic Books—Blueprints for Delinquency," *Reader's Digest,* May 1954, 24–29; C. Tilley, "Seducing the

Innocent: Fredric Wertham and the Falsifications That Helped Condemn Comics," *Information & Culture* 47 (2012): 383–413, doi:10.1353/lac.2012.0024, and D. Hajdu, *The Ten-Cent Plague: The Great Comic-Book Scare and How It Changed America* (New York: Farrar, Straus and Giroux, 2008), 8.

186 **more close relationships:** K. Hampton, L. Sessions Goulet, and K. Purcell, "Social Networking Sites and Our Lives," Pew Research Center, June 16, 2011, http://www.pewinternet.org/2011/06/16/social-networking-sites-and-our-lives/.

186 **reap psychological benefits:** E. M. Seabrook, M. L. Kern, and N. S. Rickard, "Social Networking Sites, Depression, and Anxiety: A Systematic Review," *JMIR Mental Health* 3 (2016): e50, https://doi.org/10.2196/mental5842.

186 **lead to more depression:** J. Davila et al., "Frequency and Quality of Social Networking Among Young Adults: Associations with Depressive Symptoms, Rumination, and Corumination," *Psychology of Popular Media Culture* 1 (2012): 72–86.

186 **psychological and behavioral problems:** C. Berryman, C. J. Ferguson, and C. Negy, "Social Media Use and Mental Health Among Young Adults," *Psychiatric Quarterly* 89 (2018): 307–14, and C. J. Ferguson, "Everything in Moderation: Moderate Use of Screens Unassociated with Child Behavior Problems," *Psychiatric Quarterly* 88 (2017): 797–805.

186 **troubled by insecurities:** B. A. Feinstein et al., "Negative Social Comparison on Facebook and Depressive Symptoms: Rumination as a Mechanism," *Psychology of Popular Media Culture* 2 (2013): 161–70, https://dx.doi.org/10.1037/a0033111.

186 **reviewing social-media research:** C. J. Ferguson, "The Devil Wears Stata: Thin-Ideal Media's Minimal Contribution to Our Understanding of Body Dissatisfaction and Eating Disorders," *Archives of Scientific Psychology* 6 (2018): 70–79.

186 **meta-analysis of studies:** C. J. Ferguson, "In the Eye of the Beholder: Thin-Ideal Media Affects Some, but Not Most, Viewers in a Meta-analytic Review of Body Dissatisfaction in Women and Men," *Psychology of Popular Media Culture* 2 (2013): 20–37.

186 **online social norms:** E. M. Bryant and J. Marmo, "The Rules of Facebook Friendship: A Two-Stage Examination of Interaction Rules in Close, Casual, and Acquaintance Friendships," *Journal of Social and Personal Relationships* 29 (2012): 1013–35.

186 **and users' feelings:** L. Reinecke and S. Trepte, "Authenticity and Well-Being on Social Network Sites: A Two-Wave Longitudinal Study on the Effects of Online Authenticity and the Positivity Bias in SNS Communication," *Computers in Human Behavior* 30 (2014): 95–102, and N. N. Bazarova, "Public Intimacy: Disclosure Interpretation and Social Judgments on Facebook," *Journal of Communication* 62 (2012): 815–32.

187 **most emailed list:** J. A. Berger and K. L. Milkman, "What Makes Online Content Viral?," December 5, 2009, https://ssrn.com/abstract=1528077 or https://dx.doi.org/10.2139/ssrn.1528077.

187 **measuring social buzz:** E. B. Falk, S. A. Morelli, B. L. Welborn, K. Dambacher, and M. D. Lieberman, "Creating Buzz: The Neural Correlates of Effective Message Propagation," *Psychological Science* 24 (2013): 1234–42.

187 **"While there are terrible":** Interview with Peter Sheridan Dodds, partially quoted in J. Tierney, "According to the Words, the News Is Actually Good," *New York Times,* February 23, 2015, https://www.nytimes.com/2015/02/24/science/why-we-all-sound-like-pollyannas.html.

188 **Psychology textbooks devoted:** R. F. Baumeister, C. Finkenauer, and K. D. Vohs, "Bad Is Stronger Than Good," *Review of General Psychology* 5 (2001): 324.

188 **study in 1978:** P. Brickman, D. Coates, and R. Janoff-Bulman, "Lottery Winners and Accident Victims: Is Happiness Relative?," *Journal of Personality and Social Psychology* 36 (1978): 917–27.

189 **British lottery winners:** B. Apouey and A. E. Clark, "Winning Big but Feeling No Better? The Effect of Lottery Prizes on Physical and Mental Health," *Health Economics* 24 (2015): 516–38, and J. Gardner and A. J. Oswald, "Money and Mental Wellbeing: A Longitudinal Study of Medium-Sized Lottery Wins," *Journal of Health Economics* 26 (2007): 49–60. See also A. Hedenus, "At the End of the Rainbow: Post-Winning Life Among Swedish Lottery Winners" (Ph.D. diss.,

University of Gothenburg, 2011), and J. Tierney, "How to Win the Lottery (Happily)," *New York Times,* May 27, 2014.

190 **psychologists Richard Tedeschi and Lawrence Calhoun:** R. G. Tedeschi and L. G. Calhoun, "The Posttraumatic Growth Inventory: Measuring the Positive Legacy of Trauma," *Journal of Traumatic Stress* 9 (1996): 455–71, and R. G. Tedeschi and L. G. Calhoun, "Posttraumatic Growth: Conceptual Foundations and Empirical Evidence," *Psychological Inquiry* 15 (2004): 1–18.

191 **"fading affect bias":** J. J. Skowronski, W. R. Walker, D. X. Henderson, and G. D. Bond, "The Fading Affect Bias: Its History, Its Implications, and Its Future," *Advances in Experimental Social Psychology* 49 (2014): 163–218.

192 **Those bad moments:** J. D. Green, C. Sedikides, and A. P. Gregg, "Forgotten but Not Gone: The Recall and Recognition of Self-Threatening Memories," *Journal of Experimental and Social Psychology* 44 (2008): 547–61, and J. D. Green, B. Pinter, and C. Sedikides, "Mnemic Neglect and Self-Threat: Trait Modifiability Moderates Self-Protection," *European Journal of Social Psychology* 35 (2005): 225–35.

192 **sports bettors gambled:** T. Gilovich, "Biased Evaluation and Persistence in Gambling," *Journal of Personality and Social Psychology* 44 (1983): 1110–26.

194 **This "positivity effect":** M. Mather and L. L. Carstensen, "Aging and Motivated Cognition: The Positivity Effect in Attention and Memory," *Trends in Cognitive Sciences* 9 (2005): 496–502, and A. E. Reed and L. L. Carstensen, "The Theory Behind the Age-Related Positivity Effect," *Frontiers in Psychology* 3 (2012): 339.

194 **experiment in Germany:** S. Brassen, M. Gamer, J. Peters, S. Gluth, and C. Büchel, "Don't Look Back in Anger! Responsiveness to Missed Chances in Successful and Nonsuccessful Aging," *Science* 336 (2012): 612–14.

195 **U curve of happiness:** D. G. Blanchflower and A. J. Oswald, "Is Well-Being U-shaped over the Life Cycle?," *Social Science & Medicine* 66 (2008): 1733–49; C. Graham and J. R. Pozuelo, "Happiness, Stress, and Age: How the U Curve Varies Across People and Places," *Journal of Population Economics* 30 (2017): 225–64; and J. Rauch, *The Happiness Curve* (New York: Thomas Dunne, 2018).

196 **antidepressant use, which peaks:** D. G. Blanchflower and A. J. Oswald, "Antidepressants and Age: A New Form of Evidence for U-shaped Well-Being Through Life," *Journal of Economic Behavior and Organization* 127 (2016): 46–58.

196 **chimpanzees and orangutans:** A. Weiss, J. E. King, M. Inoue-Murayama, T. Matsuzawa, and A. J. Oswald, "Evidence for a Midlife Crisis in Great Apes Consistent with the U-shape in Human Well-Being," *Proceedings of the National Academy of Sciences* 109 (2012): 19949–52.

196 **stronger immune systems:** E. K. Kalokerinos, W. von Hippel, J. D. Henry, and R. Trivers, "The Aging Positivity Effect and Immune Functions: Positivity in Recall Predicts Higher CD4 Counts and Lower CD4 Activation," *Psychology and Aging* 29 (2014): 636–41.

197 **positivity effect diminishes:** L. L. Carstensen and M. DeLiema, "The Positivity Effect: A Negativity Bias in Youth Fades with Age," *Current Opinion in Behavioral Sciences* 19 (2018): 7–12, and S. Kalenzaga, V. Lamidey, A. M. Ergis, D. Clarys, and P. Piolino, "The Positivity Bias in Aging: Motivation or Degradation?," *Emotion* 16 (2016): 602–10.

197 **At the age:** This section based on interviews with Constantine Sedikides, Tim Wildschut, Erica Hepper, Wing Yee Cheung, and Clay Routledge. For an overview of research on nostalgia, see C. Sedikides and T. Wildschut, "Finding Meaning in Nostalgia," *Review of General Psychology* 22 (2018): 48–61; C. Sedikides, T. Wildschut, J. Arndt, and C. Routledge, "Nostalgia: Past, Present, and Future," *Current Directions in Psychological Science* 17 (2008): 304–7; and J. Tierney, "What Is Nostalgia Good For? Quite a Bit, Research Shows," *New York Times*, July 8, 2013.

198 **Swiss medical dissertation:** J. Hofer, "Medical Dissertation on Nostalgia," trans. C. K. Anspach, *Bulletin of the Institute of the History of Medicine* 2 (1934): 376–91, https://www.jstor.org/stable/44437799. (Original work published in 1688.)

198 **Alpine milking song:** I. Cieraad, "Bringing Nostalgia Home: Switzerland and the Swiss Chalet," *Architecture and Culture* 6 (2018): 265–88, doi: 10.1080/20507828.2018.1477672.

198 **in other countries:** J. Beck, "When Nostalgia Was a Disease," *Atlantic*, August 14, 2013, https://www.theatlantic.com/health/archive/2013/08/when-nostalgia-was-a-disease/278648/.

199 **on cold days:** W. A. P. Van Tilburg, C. Sedikides, and T. Wildschut, "Adverse Weather Evokes Nostalgia," *Personality and Social Psychology Bulletin* 44 (2018): 984–95, doi:10.1177/0146167218756030.

199 **makes them feel warmer:** X. Zhou, T. Wildschut, C. Sedikides, X. Chen, and A. J. Vingerhoets, "Heartwarming Memories: Nostalgia Maintains Physiological Comfort," *Emotion* 12 (2012): 678–84.

199 **something bad happens:** C. Routledge, J. Arndt, C. Sedikides, and T. Wildschut, "A Blast from the Past: The Terror Management Function of Nostalgia," *Journal of Experimental Social Psychology* 44 (2008): 132–40.

199 **to counteract loneliness:** X. Zhou, C. Sedikides, T. Wildschut, and D. G. Gao, "Counteracting Loneliness: On the Restorative Function of Nostalgia," *Psychological Science* 19 (2008): 1023–29.

199 **strive toward goals:** C. Sedikides et al., "Nostalgia Motivates Pursuit of Important Goals by Increasing Meaning in Life," *European Journal of Social Psychology* 48 (2018): 209–16.

199 **stresses on the job:** M. Van Dijke, J. M. Leunissen, T. Wildschut, and C. Sedikides, "Nostalgia Promotes Intrinsic Motivation and Effort in the Presence of Low Interaction Justice," *Organizational Behavior and Human Decision Processes* 150 (2019): 46–61, doi:10.1016/j.obhdp.2018.12.003.

199 **lives have more meaning:** W. A. P. Van Tilburg, C. Sedikides, T. Wildschut, and A. J. J. Vingerhoets, "How Nostalgia Infuses Life with Meaning: From Social Connectedness to Self-Continuity," *European Journal of Social Psychology* 49 (2019): 521–32, doi:10.1002/ejsp.2519, and C. Sedikides and T. Wildschut, "Finding Meaning in Nostalgia," *Review of General Psychology* 22 (2018): 48–61, doi:10.1037/.

199 **Saul Bellow novel:** *Mr. Sammler's Planet* (New York: Penguin Classics, 2004), 156, cited in C. Sedikides and T. Wildschut, "Finding Meaning in Nostalgia," *Review of General Psychology* 22 (2018): 48–61, doi:10.1037/gpr0000109.

199 **first nostalgist, Odysseus:** E. G. Hepper, T. D. Ritchie, C. Sedikides, and T. Wildschut, "Odyssey's End: Lay Conceptions of Nostalgia Reflect Its Original Homeric Meaning," *Emotion* 12 (2012): 102–19.

200 **the magazine *Nostalgia:*** T. Wildschut, C. Sedikides, J. Arndt, and C. Routledge, "Nostalgia: Content, Triggers, Functions," *Journal of Personality and Social Psychology* 91 (2006): 975–93, and *Nostalgia Magazine,* http://www.nostalgiamagazine.net/.

200 **the more optimistic:** W. Y. Cheung, C. Sedikides, and T. Wildschut, "Induced Nostalgia Increases Optimism (via Social Connectedness and Self-Esteem) Among Individuals High, but Not Low, in Trait Nostalgia," *Personality and Individual Differences* 90 (2016): 283–88, doi: 10.1016/j.paid.20215.11.028.

200 **Martin Seligman launched:** M. Seligman, *Authentic Happiness* (New York: Simon & Schuster, 2002).

201 **exercising "socioemotional selectivity":** L. Carstensen, "The Influence of a Sense of Time on Human Development," *Science* 312 (2006): 1913–15, and L. L. Carstensen, D. M. Isaacowitz, and S. T. Charles, "Taking Time Seriously: A Theory of Socioemotional Selectivity," *American Psychologist* 54 (1999): 165–81.

202 **to "expressive writing":** S. Mugerwa and J. D. Holden, "Writing Therapy: A New Tool for General Practice?," *British Journal of General Practice* 62 (2012): 661–63.

202 **"Grief can take":** *Pudd'nhead Wilson's New Calendar,* quoted in M. Twain, *Following the Equator* (Hartford, CT: American Publishing Company, 1897), 447.

203 **from sharing joy:** S. L. Gable, H. T. Reis, E. A. Impett, and E. R. Asher, "What Do You Do When Things Go Right? The Intrapersonal and Interpersonal Benefits of Sharing Positive Events," *Journal of Personality and Social Psychology* 87 (2004): 228–45.

203 **savor an experience:** M. Biskas et al., "A Prologue to Nostalgia: Savouring Creates Nostalgic Memories That Foster Optimism," *Cognition and Emotion,* April 2, 2018, 1–11, https://doi.org/10.1080/02699931.2018.1458705.

203 **trusting and generous:** H. T. Reis et al., "Are You Happy for Me? How Sharing Positive Events with Others Provides Personal and Interpersonal Benefits," *Journal of Personality and Social Psychology* 99 (2010): 311–29.

203 **poet John Milton:** *Paradise Lost* (Oxford: Oxford University Press, 2005), bk. 5, lines 71–72.

203 **"attitude of gratitude":** A. M. Wood, S. Joseph, and J. Maltby, "Gratitude Predicts Psychological Well-Being Above the Big Five Facets," *Personality and Individual Differences* 46 (2009): 443–47.

204 **once a week:** R. A. Emmons and M. E. McCullough, "Counting Blessings Versus Burdens: An Experimental Investigation of Gratitude and Subjective Well-Being in Daily Life," *Journal of Personality and Social Psychology* 84 (2003): 377–89.

204 **the "gratitude visit":** M. E. Seligman, T. A. Steen, N. Park, and C. Peterson, "Positive Psychology Progress: Empirical Validation of Interventions," *American Psychologist* 60 (2005): 410.

205 **Stephen Stills sings:** Cited in C. Sedikides, T. Wildschut, C. Routledge, and J. Arndt, "Nostalgia Counteracts Self-Discontinuity and Restores Self-Continuity," *European Journal of Social Psychology* 45 (2015): 52–61, doi:10.1002/ejsp.2073.

205 **Humphrey Bogart dwells:** Cited in M. Vess, J. Arndt, C. Routledge, C. Sedikides, and Tim Wildschut, "Nostalgia as a Resource for the Self," *Self and Identity* 11 (2012): 273–84, http://dx.doi.org/10.1080/15298868.2010.521452.

CHAPTER 9: The Crisis Crisis

208 **"The whole aim":** H. L. Mencken, *In Defense of Women* (Mineola, NY: Dover, 2004), 29.

208 **record-store effect:** C. Morewedge, "It Was a Most Unusual Time: How Memory Bias Engenders Nostalgic Preferences," *Journal of Behavioral Decision Making* 26 (2013): 319–26.

209 **history's first economist:** M. N. Rothbard, *Economic Thought Before Adam Smith* (Cheltenham, UK: Edward Elgar, 1995).

209 **"dwelt in ease":** Hesiod, "Works and Days," lines 110–80, in *Hesiod: The Homeric Hymns and Homerica*, trans. H. G. Evelyn-White (Cambridge, MA: Harvard University Press, 1914).

210 **Slavery was an accepted tradition:** J. Black, *A Brief History of Slavery* (London: Robinson, 2011), 12–50, and M. Meltzer, *Slavery: A World History* (Boston: Da Capo, 1993). Meltzer concludes (p. 6) that the "institution of slavery was universal throughout much of history."

210 **Technological progress was slow:** Sources for the pre-eighteenth-century history include R. Stark, *How the West Won: The Neglected Story of the Triumph of Modernity* (Wilmington, DE: Intercollegiate Studies Institute, 2014), and J. Gimpel, *The Medieval Machine: The Industrial Revolution of the Middle Ages* (New York: Penguin, 1977).

210 **"eras of mankind":** Gimpel, *Medieval Machine,* viii.

211 **sway of foreign monarchs:** J. B. De Long, "Overstrong Against Thyself: War, the State, and Growth in Europe on the Eve of the Industrial Revolution," in *A Not-So-Dismal Science: A Broader View of Economies and Societies,* ed. M. Olson and S. Kähkönen (Oxford: Oxford University Press, 2004), chap. 5.

211 **the Great Enrichment:** D. N. McCloskey, *Bourgeois Equality: How Ideas, Not Capital or Institutions, Enriched the World* (Chicago: University of Chicago Press, 2016).

211 **in democratic countries:** M. Roser, "Democracy," Our World in Data, 2019, https://ourworldindata.org/democracy. See also an estimate that 72 percent of humans live in societies that are free or partly free: Freedom House, *Freedom in the World 2018,* https://freedomhouse.org/report/freedom-world/freedom-world-2018.

211 **Pinker has documented:** S. Pinker, *The Better Angels of Our Nature: Why Violence Has Declined* (New York: Penguin, 2011), Kindle, and M. Roser, "War and Peace," Our World in Data, 2019, https://ourworldindata.org/war-and-peace.

212 **farmers can feed:** J. H. Ausubel, I. K. Wernick, and P. E. Waggoner, "Peak Farmland and the Prospect for Land Sparing," *Population Development Review* 38, S1 (February 2013): 221–42, and M. Roser and H. Ritchie, "Yields and Land Use in Agriculture," Our World in Data, 2019, https://ourworldindata.org/yields-and-land-use-in-agriculture.

212 **estimated that only half:** M. Roser, "Hunger and Undernourishment," Our World in Data, 2019, https://ourworldindata.org/hunger-and-undernourishment#what-do-we-know-about-the-decline-of-under

nourishment-in-the-developing-world-over-the-long-run, citing the Food and Agriculture Organization's 50 percent estimate from the 1940s. Roser cautions that there are uncertainties about the quality of the early data but considers the estimate a "useful indication of the scale of malnutrition."

212 **nearly 90 percent:** FAO, IFAD, UNICEF, WFP, and WHO, *The State of Food Security and Nutrition in the World 2018* (Rome: FAO, 2018), 4.

212 **that life expectancy:** M. Roser, "Life Expectancy," Our World in Data, 2019, https://ourworldindata.org/life-expectancy.

212 **literacy and education:** UNESCO Institute for Statistics, "Literacy Rates—UNICEF Data," July 2018, https://data.unicef.org/topic/education/literacy/, and M. Roser, "Global Rise of Education," Our World in Data, 2019, https://ourworldindata.org/global-rise-of-education.

212 **enjoy unprecedented leisure:** J. H. Ausubel and A. Grübler, "Working Less and Living Longer: Long-Term Trends in Working Time and Time Budgets," *Technological Forecasting and Social Change* 50 (1995): 113–31, and personal communication with Jesse Ausubel, 2019. The leisure-time estimates from the 1995 paper have been updated to reflect the subsequent increase in life expectancy. For similar estimates in another country, see J. de Koning, "The Reduction of Life Hours of Work Since 1850: Estimates for Dutch Males," *SEOR Studies in Social History,* no. 2016/1, https://www.seor.nl/Cms_Media/The-shift-from-work-to-leisure.pdf.

212 **measure of human well-being:** For an overview of human progress, see Our World in Data, https://ourworldindata.org/; R. Bailey, *The End of Doom: Environmental Renewal in the Twenty-First Century* (New York: St. Martin's Press, 2015), Kindle; G. Easterbrook, *It's Better Than It Looks: Reasons for Optimism in an Age of Fear* (New York: PublicAffairs, 2018), Kindle; S. Pinker, *Enlightenment Now: The Case for Reason, Science, Humanism, and Progress* (New York: Penguin, 2018), Kindle; and S. Moore and J. Simon, *It's Getting Better All the Time: 100 Greatest Trends of the Last 100 Years* (Washington, DC: Cato Institute, 2000). For a quick look at measures of improvement, see R. Wile, "31 Charts That Will Restore Your Faith in Humanity," *Business Insider Australia,* May 23, 2013, https://www.businessinsider.com.au/charts-that-will-restore-your-faith-in-humanity-2013-5.

212 **Most respondents in:** Ipsos, *Perils of Perception*, September 2017, https://www.ipsos.com/en/global-perceptions-development-progress-perils-perceptions-research.

212 **rate of child mortality:** M. Roser, "Child Mortality," Our World in Data, 2019, https://ourworldindata.org/child-mortality.

212 **global poverty rate:** World Bank, *Poverty and Shared Prosperity 2018: Piecing Together the Poverty Puzzle* (Washington, DC: World Bank, 2018), 21.

213 **an overwhelming majority:** YouGov 2015 survey, quoted in M. Roser, "Most of Us Are Wrong About How the World Has Changed (Especially Those Who Are Pessimistic About the Future)," Our World in Data, July 27, 2018, https: ourworldindata.org/wrong-about-the-world.

213 **Gilbert and colleagues:** D. E. Levari et al., "Prevalence-Induced Concept Change in Human Judgment," *Science* 360 (2018): 1465–67.

215 **its nuclear arsenal:** G. Easterbrook, *The Progress Paradox: How Life Gets Better While People Feel Worse* (New York: Random House, 2003), Kindle.

216 **An explanation emerged:** F. K. Barlow et al., "The Contact Caveat: Negative Contact Predicts Increased Prejudice More Than Positive Contact Predicts Reduced Prejudice," *Personality and Social Psychology Bulletin* 38 (2012): 1629–43.

216 **stigmatized by the majority:** S. Paolini and K. McIntyre, "Bad Is Stronger Than Good for Stigmatized, but Not Admired Outgroups: Meta-analytical Tests of Intergroup Valence Asymmetry in Individual-to-Group Generalization Experiments," *Personality and Social Psychology Review* 23 (2019): 3–47, https://doi.org/1088868317753504.

217 **Jesse Walker documents:** J. Walker, *The United States of Paranoia: A Conspiracy Theory* (New York: HarperCollins, 2013).

217 **more than $1 trillion:** C. J. Coyne and A. Hall, "Four Decades and Counting: The Continued Failure of the War on Drugs," Cato Institute Policy Analysis No. 811, April 12, 2017, https://papers.ssrn.com/sol3/papers.cfm?abstract_id=2979445.

217 **the street prices:** D. Werb et al., "The Temporal Relationship Between Drug Supply Indicators: An Audit of International Government Surveillance Systems," *BMJ Open* 3 (2013): e003077, https://doi.org

/10.1136/bmjopen-2013-003077, and T. Bronshtein, "Explore How Illegal Drugs Have Become Cheaper and More Potent over Time," *Stat,* November 16, 2016, https://www.statnews.com/2016/11/16/illegal-drugs price potency/.

218 **fifty million Americans:** J. Dahlhamer et al., "Prevalence of Chronic Pain and High-Impact Chronic Pain Among Adults—United States, 2016," *Morbidity and Mortality Weekly Report* 67 (2018): 1001–6, http://dx.doi.org/10.15585/mmwr.mm6736a2.

218 **1 or 2 percent:** J. Sullum, "Trump Says Pain Pills Are 'So Highly Addictive.' He's Wrong," *Reason,* August 17, 2018, https://reason.com/blog/2018/08/17/trump-says-pain-pills-are-so-highly-addi, and M. Szalavitz, "Opioid Addiction Is a Huge Problem, but Pain Prescriptions Are Not the Cause," *Scientific American,* May 10, 2016, https://blogs.scientificamerican.com/mind-guest-blog/opioid-addiction-is-a-huge-problem-but-pain-prescriptions-are-not-the-cause/.

218 **medical-journal studies:** G. A. Brat et al., "Postsurgical Prescriptions for Opioid Naive Patients and Association with Overdose and Misuse: Retrospective Cohort Study," *BMJ* 360 (2018): j5790, https://www.bmj.com/content/360/bmj.j5790, and M. Noble et al., "Long-Term Opioid Management for Chronic Noncancer Pain," *Cochrane Database of Systematic Reviews* 1 (2010): CD006605, https://doi.org/10.1002/14651858.CD006605.pub2.

218 **federal government's annual:** Center for Behavioral Health Statistics and Quality, Results from the 2017 National Survey on Drug Use and Health: Detailed Tables (2018), tables 6.53B, 1.82A, 5.2A, https://www.samhsa.gov/data/sites/default/files/cbhsq-reports/NSDUHDetailedTabs2017/NSDUHDetailedTabs2017.htm#tab1-82A.

218 **The typical victim:** E. M. Johnson et al., "Unintentional Prescription Opioid-Related Overdose Deaths: Description of Decedents by Next of Kin or Best Contact, Utah, 2008–2009," *Journal of General Internal Medicine* 28 (2013): 522–29, and J. Sullum, "Opioid Commission Mistakenly Blames Pain Treatment for Drug Deaths," *Reason,* November 2, 2017, http://reason.com/blog/2017/11/02/opioid-commission-mistakenly-blames-pain.

218 **died by combining:** A. J. Visconti, G. M. Santos, N. P. Lemos, C. Burke, and P. O. Coffin, "Opioid Overdose Deaths in the City and County of

San Francisco: Prevalence, Distribution, and Disparities," *Journal of Urban Health: Bulletin of the New York Academy of Medicine* 92 (2015): 758–72; New York City Department of Health and Mental Hygiene, "Epi Data Brief," June 2016, https://www1.nyc.gov/assets/doh/downloads/pdf/epi/databrief72.pdf; and J. Sullum, "Are You More Likely to Be Killed by Opioids Than a Car Crash?," *Reason,* January 17, 2019, https://reason.com/blog/2019/01/17/are-you-more-likely-to-be-killed-by-opio. Sullum analyzes 2017 data from the CDC database Multiple Cause of Death, https://wonder.cdc.gov/mcd-icd10.html.

218 **heroin and fentanyl:** J. Bloom, "Dear PROP/CDC, Here's What Happens When You Over-Restrict Pills: More Deaths. Nice Going," American Council on Science and Health, December 12, 2018, https://www.acsh.org/news/2018/12/12/dear-propcdc-heres-what-happens-when-you-over-restrict-pills-more-deaths-nice-going-13663, and J. A. Singer, "As If We Needed It, More Evidence Emerges Showing That the Government Has Changed the Opioid Crisis into a Fentanyl Crisis," *Cato at Liberty,* November 2, 2018, https://www.cato.org/blog/we-needed-it-more-evidence-emerges-showing-government-has-changed-opioid-crisis-fentanyl-crisis.

218 **number of opioid-related deaths:** National Center for Health Statistics, "Data Brief 329," November 2018, https://www.cdc.gov/nchs/data/databriefs/db329_tables-508.pdf, analyzed in J. Sullum, "Opioid-Related Deaths Keep Rising as Pain Pill Prescriptions Fall," *Reason,* November 29, 2018, https://reason.com/blog/2018/11/29/opioid-related-deaths-keep-rising-as-pai.

218 **next Pandora's box:** J. Tierney, "The Optimists Are Right," *New York Times Magazine,* September 29, 1996, https://www.nytimes.com/1996/09/29/magazine/the-optimists-are-right.html.

219 **a violent "railway madman":** E. F. Torrey and J. Miller, *The Invisible Plague: The Rise of Mental Illness from 1750 to the Present* (New Brunswick, NJ: Rutgers University Press, 2001), 98, and J. Hayes, "The Victorian Belief That a Train Ride Could Cause Instant Insanity," *Atlas Obscura,* May 12, 2017, https://www.atlasobscura.com/articles/railway-madness-victorian-trains.

219 **Bailey has observed:** Bailey, *End of Doom,* 75.

219 **forms of demagoguery:** W. Tucker, *Progress and Privilege: America in the Age of Environmentalism* (Garden City, NY: Anchor, 1982), 89.

220 ***Our Plundered Planet:*** F. Osborn, *Our Plundered Planet* (Boston: Little, Brown, 1948).

220 ***Road to Survival:*** W. Vogt, *Road to Survival* (New York: W. Sloane Associates, 1948).

220 **Club of Rome:** D. H. Meadows, D. L. Meadows, J. Randers, and W. W. Behrens III, *The Limits to Growth* (New York: Universe Books, 1972), 23.

220 **"great die-off":** P. Ehrlich, "Looking Backward from 2000 A.D.," *Progressive,* April 1970, 23–25, quoted in V. Brent Davis, *Armageddon* (Springville, UT: CFI, 2005), 16.

220 **with "official permission":** P. R. Ehrlich, A. H. Ehrlich, and J. P. Holdren, *Ecoscience: Population, Resources, Environment* (San Francisco: W. H. Freeman, 1977), 786–77.

220 **worst human rights violations:** M. Connelly, *Fatal Misconception: The Struggle to Control World Population* (Cambridge, MA: Belknap Press of Harvard University Press, 2008), and C. Mann, *The Wizard and the Prophet: Two Remarkable Scientists and Their Dueling Visions to Shape Tomorrow's World* (New York: Knopf, 2018), 522–23.

221 **kill a billion:** P. Ehrlich, *The Machinery of Nature* (New York: Simon & Schuster, 1987), 274.

221 **2009 confirmation hearing:** U.S. Senate Committee on Commerce, Science, and Transportation, Nominations Hearing, February 12, 2009, https://www.commerce.senate.gov/public/index.cfm/hearings?ID=9BA25FEA-5F68-4211-A181-79FF35A3C6C6.

221 **less than forty thousand:** J. Hasell and M. Roser, "Famines," Our World in Data, 2019, https://ourworldindata.org/famines.

221 **to civil wars:** S. Sengupta, "Why 20 Million People Are on Brink of Famine in a 'World of Plenty,'" *New York Times,* February 22, 2017, https://www.nytimes.com/2017/02/22/world/africa/why-20-million-people-are-on-brink-of-famine-in-a-world-of-plenty.html.

221 **nearly thirty thousand predictions:** P. E. Tetlock, preface to *Expert Political Judgment* (Princeton, NJ: Princeton University Press, 2017), Kindle.

222 **most famous book:** M. Olson, *The Rise and Decline of Nations: Economic Growth, Stagflation, and Social Rigidities* (New Haven, CT: Yale University Press, 1982). Material is drawn from this book and from interviews with Olson in the 1980s.

222 **author Jonathan Rauch:** J. Rauch, *Demosclerosis: The Silent Killer of American Government* (New York: Times Books, 1994).

223 **dictum famously articulated:** G. F. Seib, "In Crisis, Opportunity for Obama," *Wall Street Journal,* November 21, 2008, https://www.wsj.com/articles/SB122721278056345271.

224 ***Crisis and Leviathan:*** R. Higgs, *Crisis and Leviathan: Critical Episodes in the Growth of American Government* (New York: Oxford University Press, 1987).

224 **stymied by Greenpeace:** D. Ropeik, "Golden Rice Opponents Should Be Held Accountable for Health Problems Linked to Vitamin A Deficiency," *Scientific American,* March 15, 2014, https://blogs.scientificamerican.com/guest-blog/golden-rice-opponents-should-be-held-accountable-for-health-problems-linked-to-vitamain-a-deficiency/.

224 **half of Americans:** Pew Research Center, "Public Perspectives on Food Risks," November 19, 2018, http://www. www.pewresearch.org/science/2018/11/19/public-perspectives-on-food-risks/.

224 **caused Zambian officials:** H. E. Cauvin, "Between Famine and Politics, Zambians Starve," *New York Times,* August 30, 2002, https://www.nytimes.com/2002/08/30/world/between-famine-and-politics-zambians-starve.html.

224 **140 Nobel laureates:** Laureates Letter Supporting Precision Agriculture (GMOs), http://supportprecisionagriculture.org/nobel-laureate-gmo-letter_rjr.html.

224 **the opinion gap:** Pew Research Center, "Major Gaps Between the Public, Scientists on Key Issues," July 1, 2015, http://www.pewinternet.org/interactives/public-scientists-opinion-gap/.

225 **now $3 billion:** J. A. DiMasi, H. G. Grabowski, and R. W. Hansen, "Innovation in the Pharmaceutical Industry: New Estimates of R&D Costs," *Journal of Health Economics* 47 (2016): 20–33, and personal communication with DiMasi, who updated his calculations in 2018.

225 **at $6 billion:** A. S. A. Roy, "Stifling New Cures: The True Cost of Lengthy Clinical Drug Trials," Manhattan Institute for Policy Research, Project FDA Report 5 (April 2012).

226 **Nicotine in itself:** Royal Society for Public Health, "Nicotine 'No More Harmful to Health Than Caffeine,'" August 13, 2015, https://www.rsph.org.uk/about-us/news/nicotine—no-more-harmful-to-health-than-caffeine-.html. The Royal College of Physicians similarly concluded, after reviewing the literature, that nicotine "itself is not especially hazardous"; see Tobacco Advisory Group of the Royal College of Physicians, *Harm Reduction in Nicotine Addiction: Helping People Who Can't Quit* (London: RCP, 2007), preface.

226 **Nicotine has been shown:** S. J. Heishman, B. A. Kleykamp, and E. G. Singleton, "Meta-analysis of the Acute Effects of Nicotine and Smoking on Human Performance," *Psychopharmacology* 210 (2010): 453–69.

226 **15 percent of adults:** European Commission, "Attitudes of Europeans Towards Tobacco and Electronic Cigarettes," Special Eurobarometer 458, March 2017, table QB1, https://ec.europa.eu/commfrontoffice/publicopinion/index.cfm/ResultDoc/download/DocumentKy/79002.

226 **that 350,000 lives:** A. Milton, C. Bellander, G. Johnsson, and K. O. Fagerström, "Snus Saves Lives: A Study of Snus and Tobacco-Related Mortality in the EU," Snus Commission's Third Report, May 2017, https://www.clivebates.com/documents/SnusCommissionJune2017.pdf.

226 **Baptist-bootlegger coalition:** J. H. Adler, R. E. Meiners, A. P. Morriss, and B. Yandle, "Baptists, Bootleggers & Electronic Cigarettes," *Yale Journal on Regulation* 33 (2016): 313–61, http://digitalcommons.law.yale.edu/yjreg/vol33/iss2/1.

227 **studies of snus:** P. N. Lee and J. Hamling, "Systematic Review of the Relation Between Smokeless Tobacco and Cancer in Europe and North America," *BMC Medicine* 7 (2009): 36, https://doi.org/10.1186/1741-7015-7-36.

227 **kept snus illegal:** G. Ross, "The EU's New Tobacco Directive: Protecting Cigarette Markets, Killing Smokers," *Forbes,* January 10, 2013, https://www.forbes.com/sites/realspin/2013/01/10/the-eus-new-tobacco-directive-protecting-cigarette-markets-killing-smokers/.

227 **prevented snus merchants:** H. Campbell, "FDA Denies Modified Risk Tobacco Product Status for Snus," American Council on Science and Health, December 14, 2016, https://www.acsh.org/news/2016/12/14/fda-denies-modified-risk-tobacco-product-status-snus-10592.

227 **rigorous study in 2019:** P. Hajek et al., "A Randomized Trial of E-Cigarettes Versus Nicotine-Replacement Therapy," *New England Journal of Medicine* 380 (2019): 629–37, https://doi.org/10.1056/NEJMoa1808779.

227 **below 15 percent:** T. W. Wang et al., "Tobacco Product Use Among Adults—United States, 2017," *Morbidity and Mortality Weekly Report* 67 (2018): 1225–32, http://dx.doi.org/10.15585/mmwr.mm6744a2, and National Center for Health Statistics, "Summary Health Statistics: National Health Interview Survey, 2017," table A-12a, https://ftp.cdc.gov/pub/Health_Statistics/NCHS/NHIS/SHS/2017_SHS_Table_A-12.pdf. For the historical trend, see A. LaVito, "CDC Says Smoking Rates Fall to Record Low in US," CNBC, November 8, 2018, https://www.cnbc.com/2018/11/08/cdc-says-smoking-rates-fall-to-record-low-in-us.html.

227 **vaping stick, the Juul:** J. Tierney, "Juul Madness," *City Journal,* July 15, 2018, https://www.city-journal.org/html/juul-labs-vaping-prohibitionists-16029.html.

227 **outlaw or severely restrict:** J. Tierney, "The Corruption of Public Health," *City Journal,* Summer 2017, https://www.city-journal.org/html/corruption-public-health-15323.html, and J. Sullum, "Surgeon General Undermines Harm Reduction by Pushing Anti-vaping Policies and Propaganda," *Reason,* December 18, 2018, https://reason.com/blog/2018/12/18/surgeon-general-undermines-harm-reductio.

228 **95 percent safer:** Royal College of Physicians, *Nicotine Without Smoke: Tobacco Harm Reduction* (London: RCP, 2016), 58, 185. A similar conclusion was reached by England's national health agency in A. McNeill et al., "E-Cigarettes: An Evidence Update," a report commissioned by Public Health England, 2015, 80. See also D. J. Nutt et al., "Estimating the Harms of Nicotine-Containing Products Using the MCDA Approach," *European Addiction Research* 20 (2014): 218–25, https://doi.org/10.1159/000360220, and M. Siegel, "National Academy of Sciences Report on Electronic Cigarettes Confirms That Vaping Is

Much Safer Than Smoking and Has No Known Long-Term Health Effects," *The Rest of the Story: Tobacco and Alcohol News Analysis and Commentary,* January 24, 2018, http://tobaccoanalysis.blogspot.com /2018/01/national-academy-of-sciences-report-on_24.html.

228 **to historic lows:** D. T. Levy et al., "Examining the Relationship of Vaping to Smoking Initiation Among US Youth and Young Adults: A Reality Check," *Tobacco Control,* November 20, 2018, https://doi.org /10.1136/tobaccocontrol-2018-054446, and B. Rodu, "Federal Officials, Please Pay Attention to Federal Surveys: E-Cigarettes Are Not Gateway Products," *Tobacco Truth,* February 21, 2018, https://rodutobaccotruth .blogspot.com/2018/02/federal-officials-please-pay-attention.html.

228 **majority of teenagers:** B. Rodu, "The FDA's Teen E-Cigarette-Addiction Epidemic Doesn't Add Up," *Tobacco Truth,* October 1, 2018, https:// rodutobaccotruth.blogspot.com/2018/10/the-fdas -teen-e-cigarette-addiction.html, analyzing data from the CDC's National Youth Tobacco Survey, 2017, https://www.cdc.gov/tobacco/data _statistics/surveys/nyts/index.htm.

228 **more teenage drinkers:** National Institute on Drug Abuse, "Monitoring the Future Survey: High School and Youth Trends," December 2018, https://www.drugabuse.gov/publications/drugfacts/monitoring -future-survey-high-school-youth-trends.

228 **cut in half:** The CDC and the FDA have worked hard to divert attention from the good news about youth smoking rates. They have repeatedly issued press releases and bulletins about alarming increases in youths' "tobacco use"—a misleading term because it includes vaping sticks along with tobacco cigarettes—while offering the public little information about the downward cigarette-smoking trends in federal usage surveys. But the data for students from the CDC's annual National Youth Tobacco Survey is aggregated in one CDC fact sheet ("Youth and Tobacco Use," https://www.cdc.gov/tobacco/data_statis tics/fact_sheets/youth_data/tobacco_use/index.htm), which shows that from 2011 to 2018, the rate of current cigarette smoking declined among high-school students to 8.1 percent from 15.8 percent, and among middle-school students to 1.8 percent from 4.3 percent. The downward trend among young adults has been similar, according to data from the CDC's National Health Interview Survey that was

aggregated in testimony by Bill Godshall on January 18, 2019, to the FDA hearing "Eliminating Youth Electronic Cigarette Use: The Role for Drug Therapies." The CDC data showed that from 2010 to 2017, the rate of current smoking among adults aged eighteen to twenty-four declined to 10.4 percent from 20.1 percent.

229 **reversal in popular opinion:** J. Huang et al., "Changing Perceptions of Harm of e-Cigarette vs Cigarette Use Among Adults in 2 US National Surveys from 2012 to 2017," *JAMA Network Open* 2 (2019): e191047, doi:10.1001/jamanetworkopen.2019.1047. A more recent national survey found that only 20 percent of American adults judged electronic cigarettes to be safer than tobacco cigarettes, while 50 percent believed them to be just as harmful, and 13 percent considered them more harmful. See Rasmussen Reports, "Most Say E-Cigarettes No Healthier Than Traditional Ones," August 16, 2018, http://www.rasmussenreports.com/public_content/lifestyle/general_lifestyle/august_2018/most_say_e_cigarettes_no_healthier_than_traditional_ones.

229 **Europeans are similarly misinformed:** European Commission, "Attitudes of Europeans Towards Tobacco and Electronic Cigarettes," Special Eurobarometer 458, March 2017, p. 22, https://ec.europa.eu/commfrontoffice/publicopinion/index.cfm/ResultDoc/download/DocumentKy/79002.

229 **the Optimism Gap:** D. Whitman, *The Optimism Gap: The I'm OK—They're Not Syndrome and the Myth of American Decline* (New York: Walker, 1998).

230 **pessimistic about the economy:** B. S. Bernanke, "How Do People Really Feel About the Economy?," *Brookings,* June 30, 2016, https://www.brookings.edu/blog/ben-bernanke/2016/06/30/how-do-people-really-feel-about-the-economy/.

230 **to television news:** M. E. McNaughton-Cassill and T. Smith, "My World Is OK, but Yours Is Not: Television News, the Optimism Gap, and Stress," *Stress and Health* 18 (2002): 27–33.

230 **the richest counties:** T. P. Jeffrey, "Census Bureau: 5 Richest U.S. Counties Are D.C. Suburbs; 10 of Nation's 20 Richest Counties in D.C. Area," CNS News, December 6, 2018, https://www.cnsnews.com/news/article/terence-p-jeffrey/census-bureau-5-richest-counties-still-dc-suburbs-10-top-20.

231 **an existential threat:** I. Schwartz, "Obama: Like I Said at the Baseball Game, ISIS 'Can't Destroy Us,' 'They Are Not an Existential Threat,'" RealClearPolitics, March 23, 2016, https://www.realclearpolitics.com/video/2016/03/23/obama_like_i_said_at_the_baseball_game_isis_cant_destroy_us_they_are_not_an_existential_threat.html.

231 **ranks much lower:** B. Lomborg, *Global Problems, Smart Solutions* (New York: Cambridge University Press, 2014).

232 **1970s and 1980s:** C. York, "Islamic State Terrorism Is Serious but We've Faced Even Deadlier Threats in the Past," *Huffington Post,* November 29, 2015, https://www.huffingtonpost.co.uk/2015/11/28/islamic-state-terrorism-threat_n_8670458.html.

232 **polar bears going extinct:** S. J. Crockford, *State of the Polar Bear Report 2018* (London: Global Warming Policy Foundation, 2018), https://www.thegwpf.org/content/uploads/2018/02/Polarbears2017.pdf; M. Ridley, "The Polar Bear Problem," *The Rational Optimist,* August 11, 2011, http://www.rationaloptimist.com/blog/the-polar-bear-problem/; and J. Tierney, "The Good News Bears," *New York Times,* August 6, 2005, https://www.nytimes.com/2005/08/06/opinion/the-good-news-bears.html.

232 **Enact Patty's Law:** J. Bleyer, "Patty Wetterling Questions Sex Offender Laws," *City Pages,* March 20, 2013, http://www.citypages.com/news/patty-wetterling-questions-sex-offender-laws-6766534, and T. Frank, "If a Law Has a First Name, That's a Bad Sign," *Los Angeles Times,* September 19, 2016, http://www.latimes.com/opinion/op-ed/la-oe-frank-named-laws-20160919-snap-story.html.

234 **red team exercise:** M. Zenko, *Red Team: How to Succeed by Thinking Like the Enemy* (New York: Basic Books, 2015), and S. Koonin, "A 'Red Team' Exercise Would Strengthen Climate Science," *Wall Street Journal,* April 20, 2017, https://www.wsj.com/articles/a-red-team-exercise-would-strengthen-climate-science-1492728579.

234 **the journal *Science*:** J. Simon, "Resources, Population, Environment: An Oversupply of False Bad News," *Science* 208 (1980): 1431–37.

235 **bet came due:** J. Tierney, "Betting on the Planet," *New York Times Magazine,* December 2, 1990, https://www.nytimes.com/1990/12/02/magazine/betting-on-the-planet.html.

235 **winnings in 2010:** J. Tierney, "Economic Optimism? Yes, I'll Take That Bet," *New York Times,* December 27, 2010.

235 **at Long Bets:** Long Bets: The Arena for Accountable Predictions, longbets.org, and J. Tierney, "Can Humanity Survive? Want to Bet on It?," *New York Times,* January 30, 2007, https://www.nytimes.com/2007/01/30/science/30tier.html.

235 **by Warren Buffett:** A. Kabil, "How Warren Buffett Won His Multi-Million Dollar Long Bet," *Medium,* February 17, 2018, https://medium.com/the-long-now-foundation/how-warren-buffett-won-his-multi-million-dollar-long-bet-3af05cf4a42d.

236 **a "media contagion":** S. Towers, A. Gomez-Lievano, M. Khan, A. Mubayi, and C. Castillo-Chavez, "Contagion in Mass Killings and School Shootings," *PLoS ONE* 10 (2015): e0117259, https://doi.org/10.1371/journal.pone.0117259.

236 **couple of organizations:** Don't Name Them, http://www.dontnamethem.org/, and No Notoriety, https://nonotoriety.com/.

237 **A few journalists:** B. Stelter, "'No Notoriety' Campaign to Not Name Mass Murderers Sees Progress," CNN Business, October 2, 2015, https://money.cnn.com/2015/10/02/media/media-decisions-naming-showing-killers/index.html, and M. Follman, "How the Media Inspires Mass Shooters, and 7 Ways News Outlets Can Help Prevent Copycat Attacks," *Mother Jones,* October 6, 2015, https://www.motherjones.com/politics/2015/10/media-inspires-mass-shooters-copycats/.

CHAPTER 10: The Future of Good

241 **ability to memorize:** Plato, *Plato's Phaedrus: Complete & Unabridged Jowett Translation,* trans. B. Jowett (Boston: Actonian Press, 2010), section 275, Kindle.

241 **Ottoman Empire banned:** M. M. Coşgel, T. J. Miceli, and J. Rubin, "Guns and Books: Legitimacy, Revolt and Technological Change," University of Connecticut Department of Economics Working Paper Series 2009–12, March 2009, https://opencommons.uconn.edu/cgi

/viewcontent.cgi?article=1256&context=econ_wpapers, and E. B. Ekinci, "Myths and Reality About the Printing Press in the Ottoman Empire," *Daily Sabah,* June 8, 2015, https://www.dailysabah.com/feature/2015/06/08/myths-and-reality-about-the-printing-press-in-the-ottoman-empire.

241 **banned heretical books:** M. Lenard, "On the Origin, Development and Demise of the *Index Librorum Prohibitorum,*" *Journal of Access Services* 3 (2006): 51–63, https://doi.org/10.1300/J204v03n04_05.

241 **Newspapers were blamed:** V. Bell, "Don't Touch That Dial!," *Slate,* February 15, 2010, https://slate.com/technology/2010/02/a-history-of-media-technology-scares-from-the-printing-press-to-facebook.html.

242 **professor of "media ecology":** N. Postman, *Technopoly: The Surrender of Culture to Technology* (New York: Knopf, 1992), and N. Postman, *Amusing Ourselves to Death: Public Discourse in the Age of Show Business* (New York: Viking, 1985).

242 **like *Data Smog:*** D. Shenk, *Data Smog: Surviving the Information Glut* (New York: HarperOne, 1997).

242 ***Life After Television:*** G. Gilder, *Life After Television: The Coming Transformation of Media and American Life* (New York: W. W. Norton, 1992).

242 **Gilder said in 1997:** J. Tierney, "Technology Makes Us Better; Our Oldest Computer, Upgraded," *New York Times Magazine,* September 28, 1997, https://www.nytimes.com/1997/09/28/magazine/technology-makes-us-better-our-oldest-computer-upgraded.html.

243 **polarization of opinion:** M. P. Fiorina, *Unstable Majorities: Polarization, Party Sorting & Political Stalemate* (Stanford, CA: Hoover Institution Press, 2017); L. Boxell, M. Gentzkow, and J. M. Shapiro, "Greater Internet Use Is Not Associated with Faster Growth in Political Polarization Among US Demographic Groups," *Proceedings of the National Academy of Sciences* 114 (2017): 10612–17, https://doi.org/10.1073/pnas.1706588114; S. Iyengar, G. Sood, and Y. Lelkes, "Affect, Not Ideology: A Social Identity Perspective on Polarization," *Public Opinion Quarterly* 76 (2012): 405–31, https://doi.org/10.1093/poq/nfs038; and Pew Research Center, "Trends in American Values: 1987–2012," June 4, 2012, http://www.people-press.org/2012/06/04/partisan-polarization-surges-in-bush-obama-years/.

244 **government's dubious advice:** G. Taubes, *Good Calories, Bad Calories: Fats, Carbs, and the Controversial Science of Diet and Health* (New York: Knopf, 2007), and N. Teicholz, *The Big Fat Surprise: Why Butter, Meat, and Cheese Belong in a Healthy Diet* (New York: Simon & Schuster, 2014).

244 **changing their diets:** C. D. Rehm, J. L. Peñalvo, A. Afshin, and D. Mozaffarian, "Dietary Intake Among US Adults, 1999–2012," *JAMA* 315 (2016): 2542–53, https://doi.org/10.1001/jama.2016.7491, and M. Sanger-Katz, "Americans Are Finally Eating Less," *New York Times,* July 24, 2015, https://www.nytimes.com/2015/07/25/upshot/americans-are-finally-eating-less.html.

244 **book on self-control:** R. F. Baumeister and J. Tierney, *Willpower: Rediscovering the Greatest Human Strength* (New York: Penguin Press, 2011): 215–37.

246 **"Men, it has":** Charles Mackay, *Extraordinary Popular Delusions and the Madness of Crowds* (Lexington, KY: Seven Treasures, 2008), "National Delusions," Kindle.

Index

Page numbers above 250 refer to the endnotes.